A SHEARWATER BOOK

PLACE OF THE WILD

A Wildlands Anthology

PLACE
of the
WILD

Edited by

David Clarke Burks

ISLAND PRESS / Shearwater Books

Washington, D.C. / *Covelo, California*

A Shearwater Book
published by Island Press

Copyright © 1994 Island Press
All rights reserved under International and Pan-American Copyright
Conventions. No part of this book may be reproduced in any form
or by any means without permission in writing from the publisher:
Island Press, 1718 Connecticut Avenue, N.W., Suite 300, Washington,
D.C. 20009.
Shearwater Books is a trademark of The Center for Resource Economics.

Acknowledgment for permission to include previously published
material is expressed on p. 330.

LIBRARY OF CONGRESS CATALOGING-IN-PUBLICATION DATA
Place of the wild : a wildlands anthology /edited by David Clarke
Burks.
   p. cm.
   Includes bibliographical references (p.   ) and index.
   ISBN 1-55963-341-7 (cloth : acid-free paper). — ISBN
1-55963-342-5 (paper : acid-free paper)
    1. Nature conservation.   2. Wilderness areas.   3. Human ecology.
I. Burks, David Clarke.
QH75.P493 1994
333.7'2—dc20
                                   94-29221
                                       CIP

Printed on recycled, acid-free paper

Manufactured in the United States of America

10 9 8 7 6 5 4 3 2

FOR THE WOLVES AND THE BEARS

AND THE RED-BACKED VOLES,

AND ALL THE FOLKS WORKING

TO MAKE THE WILDLANDS PROJECT VISION A REALITY

Contents

A properly radical environmentalist position is in no way anti-human. We grasp the pain of the human condition in its full complexity, and add the awareness of how desperately endangered certain key species and habitats have become. . . . The critical argument now within environmental circles is between those who operate from a human-centered resource management mentality and those whose values reflect an awareness of the whole of nature.

GARY SNYDER, *The Practice of the Wild*

Why ought man to value himself as more than an infinitely small composing unit of the one great unit of creation? . . . The universe would be incomplete without man, but it would also be incomplete without the smallest transmicroscopic creature that dwells beyond our conceitful eyes and knowledge.

JOHN MUIR, *A Thousand-Mile Walk to the Gulf*

PLACE

of the

WILD

INTRODUCTION

Raven, the trickster, was tired of living in darkness.
He could not see the beauty of the world around him.
Raven was determined to find a way to bring the sun's light
into the world from its hiding place in the house of the old man.

—*John Swanton, adapted from* Tlingit Myths and Texts

Contours of the Wild

DAVID CLARKE BURKS

The first thing about rocks is, they're old. . . . Rocks are in time in a different way than living things are, even the ancient trees. But then, the other thing about rocks is that they are place. Rocks are what a place is made of to start with and after all. . . . The stone is at the center.

URSULA LE GUIN, *Buffalo Gals and Other Animal Presences*

Where and what *is* the place of the wild? There are as many answers to this compound question as there are defenders of wilderness, and the narratives are as diverse as the assembly of life in an old-growth forest. This collection of new writings on wildlands and wilderness explores the varied contours of the wildlands countryside and provides readers a feel for the textures and pitfalls of the terrain. There is no single compass point or prominent peak for reference along the trail. Discovering a way into and through the place of the wild is always a personal journey. What I offer here is some background, a preview of what you are likely to come across, and some signposts to guide your way through these pages.

This work is not a dispassionate monograph designed to sit alongside other reductionist treatises on a library shelf. Each of the authors is an advocate for the preservation and restoration of wildlands and wilderness. Some are outspoken, unhousebroken, unabashed activists whose daily meat is the protection of wild places and wild species. Some are nesters at home with the work and the stories of their place. Others are ecophilosophers engaged in the construction of frameworks and paradigms of critical discovery. And still others fall outside any scheme of classification because they take their place in the only way they can—by living it one word and one deed at a time.

Every book, like every living species, has its origins in time and place. This one grew out of a soup of recombinant dialogues between three Davids (Foreman, Johns, and Burks) in early 1993. Among this group of wildlands proponents a nagging question kept coming to the fore: What should be the appropriate role of humans vis-à-vis natural habitats? (See John Davis's essay for background on The Wildlands Project.) Framed another way: Is contemporary Western culture compatible with the goal of preserving biodiversity across the landscape of North America? Put still another way: What do you do about humans who seem to get their meddling fingers into everything and most often spoil the soup? A decision was made to put these questions to a group of people who care deeply about this issue, and this book is the result.

Each of the writers was invited to contribute a new or substantially revised piece for the anthology. What you find here is reflective of each contributor's most current work. Any underlying epistemological coherence owes more to happenstance than to design. It falls to the reader to discover the patterns which weave among these narratives, poems, and essays. Clearly each piece is grounded in the author's personal experience of the wild. As David Abram writes in his essay, "to tell the story . . . is to let the place itself speak through the telling."

The common thread which draws these essays together is a shared conviction that academics and activists, philosophers and poets, must join forces to begin reweaving and recreating the cultural narratives and shared visions that work to support the physical processes which sustain life on this planet. There is an uneasy recognition that should we fail to maintain wildlands and wilderness, we will lose access not only to the natural world but to that original place of the human spirit. William Kittredge writes in another forum: "Our most urgent social and political question [is]: how to live in right relationship. In learning to pay respectful attention to one another and plants and animals, we relearn the acts of empathy, and thus humility and compassion— ways of proceeding that grow more and more necessary as the world crowds in."

Over the past decade there has been a marked shift in the character of the debates over wilderness preservation. The parameters have expanded. The forward edge of the debate has reached beyond an immediate goal of protecting unique enclaves of natural scenery to more expansive prospects of incorporating vast reserves of biological diversity into a network of wildlands, refugia, and wilderness areas across North America (Turtle Island). The theme here is the preservation of wilderness itself, for itself, not merely for human recreation.

Barry Lopez, a fellow dweller in the land of tall trees west of the Cascades, has ably summarized some key components of the current wilderness preservation position in an essay entitled "Yukon Charley: The Shape of Wilderness":

An argument for wilderness that reaches beyond the valid concerns of multiple use—recreation, flood control, providing a source of pure water—is that wildlands preserve complex biological relationships that we are only dimly, and sometimes not at all aware of. Wilderness represents a gene pool vital for the resiliency of plants and animals. An argument for wilderness that goes deeper still is that we have an ethical obligation to provide animals with a place where they are free from the impingements of civilization. And, further, an historical responsibility to preserve the kind of landscapes from which modern man emerged.

The survival of wildlands, where the dominant forces are not human but native and biologically diverse, is contingent upon reaching an accord between the forces of culture and the imperatives of biology, as Ed Grumbine illustrates in his narrative. It is apparent that human actions, especially in recent times, have radically altered natural patterns on the landscape. It is also quite apparent that human interventions have degraded and imperiled natural processes that are critical to sustain healthy, diverse ecosystems. It is less apparent, but equally true, that these interventions have eroded the character of human community and our sense of place.

To those used to the familiar patter of conventionally bland environmental rhetoric, a few of these essays may appear strident. I suppose some of them are. But where have conventional rhetoric and compromise taken us, except to the edge of species extinction? A number of these writers are asking us to crawl out on the ledge with them and peer over. They are telling us it is time we took a risk to see what this world could be like if we set aside our compulsive urge to control all of nature.

The patterns of the wild are often at odds with the square grids of culture. It is not enough to make and enforce laws; we must discover the place of the wild in our own lives. As Stephanie Mills writes in her essay, "for humans to locate themselves in right relationship with wild nature . . . will require a radical rededication to the wild."

Nothing less than revisioning human relationships with the wild and wildlands is called for. Somewhat akin to the debate which raged following Frederick Jackson Turner's essay one hundred years ago positing the end of the American frontier, the current debate engages the consequences of diminishing wildlands across North America. The effects of human infringement on wild places are well documented. In a world becoming tamed, flattened, and abraded under the pile driver of Western technology, we must recognize that the relatively recent human experiment may go the way of the ground sloth and the saber-toothed tiger. As E. O. Wilson writes in *The Diversity of Life*, "the assembly of life that took a billion years to evolve, all this up until recently has seemed timeless, immutable, but now the destructive actions of humans have forced us to pose the question: how much force does it take to break the crucible of evolution?"

Preserving the place of the wild, in both the geography and consciousness of North America, suggests a need for limits and self-restraint. You will hear voices in this collection echoing that theme. Limits on individual action are an especially provocative subject in a culture committed to the cult of individualism. See particularly the essays by Dave Foreman and Bill McKibben. And no one doubts that current population pressures will make the preservation of wild places extremely difficult in the decades and centuries ahead. Monique Miller offers some compelling perspectives on population trends and consequences.

Each generation rewrites the past, and as we bring the concerns of the present to bear on the past, we would be wise to remember the concerns of the past as they bear on the present. A number of writers whose works fall within the "Frameworks" section of this anthology address the origins of our current dilemmas. They cite the works of scholars and teachers, of natural history writers and anthropologists, who rummage through the middens of human history to uncover the origins and consequences of human interventions. Without new frameworks and metaphors to organize our perceptions, we are fated to repeat our mistakings.

One of the most provocative new frameworks is provided by researchers of chaos and complexity. Jack Turner explores some of the challenges to traditional reductionist thinking posed by chaos theory. Max Oelschlaeger in this same section takes us on a journey through wilderness as a deep ecological ethic. Paul Faulstich, on the other hand, enjoins us to understand wilderness as a human construct. There is some disparity between the ideas of Faulstich and those of Foreman. And that is to be expected and welcomed as defenders of wildlands push against the boundaries of current, politically correct, reservoirs of human knowledge about ourselves. The harder we push, the more the dam weakens and crumbles so that the river can once again assume its natural flow.

Like proponents of the new science of chaos, wildlands advocates have an eye for pattern but also a taste for randomness—for jagged edges and sudden leaps. At any point in a chain of events a crisis, such as a wildfire, may send off magnified signals to distant points in the system. The consequences of these sudden fluctuations are rarely predictable. The voices in this collection are such wildfires. In some instances their admonitions may appear discontinuous and erratic when measured against traditional wilderness doctrines and deterministic calls for "ecosystem management." But as we enter their experiences and look closer at the place of the wild, we begin to recognize shapes, paths, clusterings, and tangles as organizing principles and start to understand wilderness as a process of becoming. It is here we begin to make connections out of seeming disorder—connections between the natural world and the human mind.

There is a rich diversity in this collection of essays. You can take them apart and put them together again and they do not add up to a single viewpoint. Like the weather and the clouds, their movements are not predictable but stay within certain bounds. They describe a distinctive shape, a kind of double spiral in three dimensions, like a butterfly. From chaos theory comes a metaphor called the Butterfly Effect. It states: a butterfly stirring the air today in Beijing can transform the weather next month in New York. The voices here are signaling the shape of a new kind of order for wildlands and humans in the centuries ahead.

———

It has been in relation to the place of the wild that the human mind has developed its powers of reflective thought. In a work entitled *Biophilia*, E. O. Wilson has written: "Something moves on the edge of a field of vision, a new connection is glimpsed, holds for a moment. Words pour in and around, and the image takes substantial form, at first believed familiar, then seen as strikingly new." Wilson goes on to remind us we are in the fullest sense a biological species and will find no ultimate meaning apart from the other life on this planet.

It is here we recognize that the wild, epitomized by biologically diverse wildlands and wilderness, is crucial not only to the health and welfare of ecosystems but to human ontogeny and well-being also. You will find these notions given flesh in a number of essays including those by Alan Drengson and Dolores LaChapelle. Drengson looks for "the indigenous wisdom of place" in the midst of a one-million-acre wilderness. And LaChapelle follows the connections between humans and place by exploring "the vast interlocking rhizomes [that] link each to all." Gary Nabhan, through his well-crafted narrative, explains what it means to inhabit indigenous relationships with place. It is more than understanding that all life is connected; it is acting in that awareness, doing the work of connection, subordinating human will to the will of place.

Both culture and wildlands are, in one sense, metaphors constructed to prefigure the relations between self and other. Analogues and images, Paul Shepard reminds us in his works on human ecology, are the gateways by which our minds explore and map terrain beyond the self. We build memory by linking new concepts to old ones. Both Nancy Lord and John Haines, two notable Alaskan writers, point out in their narratives the deep connections shaped by living close to the wild. The mind and the body probe and adapt to changes in the landscape and changes in the seasons, and always the search continues for analogues to place in memory.

II

The theme of connection is pervasive throughout the anthology. Another contributor, Sierran poet and writer Gary Snyder, has written elsewhere in *The Practice of the Wild*: "To know the spirit of a place is to realize that you are part of a part and that the whole is made of parts, each of which is whole. You start with the part you are whole in." The work of preserving and restoring wildlands and wilderness is contingent on humans learning to live appropriately in place. Bill Devall's account of forest restoration work in Redwood National Park is an excellent example of getting to know place. And Mollie Matteson's essay on the extirpation of wolves in the Judith Basin of Montana warns what it means to erase the last vestiges of wildness. Doing one's homework is a precondition for working one's home. Recovery is a process of discovery—of claiming territory, not in the acquisitive sense of personal property, but claiming one's place in place, as a moral and ecological constituent of the whole.

I am brought home to the resonance of my place in the Willamette Valley west of the Cascades as I walk the broad shoulders of nearby Mount Pisgah. Ascending through forests of Douglas fir and cedar, I come at last to an upland meadow populated by Oregon white oaks and a scattering of blue camas flowers. Red-tailed harriers soar in the thermals above the summit while chickadees and Black-headed Grosbeaks flick through the underbrush of swordfern and salal. I walk and listen and inhale the untamed patterns of this place. All of us have narratives of connection with place. One can even unearth those connections in a place like New York City, as Terry Tempest Williams and Margaret Hays Young underscore in their essays. Through an awareness of our being in the world we come to understand that wildness is both an outward fact and a quality of experience. It is out of these personal efforts to incorporate the wild that we inform our understanding of the "big outside" and begin to feel the importance of the place of the wild in ourselves.

———

How we think about the wild, and whether we think about it at all, depends on our proximity to that place in our minds "where the wild things are." For all but a few, the tamed, the managed, and the controlled are our daily fields of reference. Our thoughts and actions are sifted through the grid of contemporary urban consciousness. The wild is, therefore, extraordinary, implausible, and often inaccessible to that part of our brain allotted to reason and rationality. But more than a few intrepid explorers have come to believe that another, older brain still works inside us, a response system still attuned to native ecology. When we touch that place, we trigger a host of images, reflexes, cognitions which remind us of our coevolutionary history with plants and animals in the wild. Culture overlays our perceptions, but it does not erase our origins.

Is it possible to think and act outside the paradigms laid down by culture? This is a question which has occupied philosophers and social scientists for generations. I suspect the answer is no. We cannot just discard the lenses which have been ground to a particular prescription by history, accident, and adaptation. But cultures are varied and numerous. There are many myths and texts to sift through. Perhaps we can learn to incorporate sustainable narratives into the ground of our lives. If so, we will need to work to marry story with science, natural history with narrative. In the selections from poet Alison Deming's work *The Monarchs*, we are urged "to listen hard" to the wild biology that surrounds us "and figure out / from scratch how to stay alive."

Our minds work in narrative; relationships are clarified and tested in story. The contributors to this anthology are builders of frameworks, architects of place, poets of relationship. Each of them in her or his own life is attempting to understand the place of the wild. Knowledge of *who* we are is embedded in the soil of *where* we are. Wildlands are maps which connect us to the voice of the earth. And recovery of wild ecosystems goes to the very heartwood of the matter. Without human efforts to preserve and restore vast, interconnected reserves for the unimpeded work of evolution and biological succession, the very survival of the web that supports life is imperiled. Participating in restoration work is a means of becoming integral to land and place, not as abstract "environment" but as shared habitat. To protect the wild, humans will have to forgo the impulse to dominate land as a resource. We will need to curb our appetites in order to assure protection for at least two types of places: one where human habitation leaves few traces; and the other where people operate as faithful stewards who inhabit and consume within sustainable limits the interest, not the capital, of the land.

One can hear, rustling in the pages of this book, the stirrings of a new wilderness movement—one committed to the restoration of wild landscapes and biomes that reach far beyond our current islands of "rock and ice." It is a movement working ardently to reset the terms of the debate over wilderness. Grounded in commitment to biological diversity and informed by principles from the new science of conservation biology, it is seeking to foment changes in the very map of human relationships with the place of the wild. Uncovering the links which connect the human spirit with its source in place will begin to lay the groundwork for a new land ethic in the twenty-first century. The writings in this collection are no less than such an undertaking.

A book, however, is always and only an artifact—an assemblage of words and metaphors that take on significance as they become incorporated into our lives and works. The voices found between the covers of this anthology are evocations. They call us to recognize an umbilical connection to more than our own kind, to spread our toes in the thick loam of nations—plant and animal nations and all manner of families, clans, and moieties from Raven to defenders of wilderness. It falls to each of us to engage in the work of recovery and to discover the place of the wild, both outside and inside ourselves.

A NOTE ON THE ILLUSTRATIONS

Northwest Coast Native American design motifs were selected to complement and inform the written contributions to the anthology. The four pieces presented here depicting Raven, Salmon, Bear, and Wolf owe much to the work of the eminent Haida artist Robert Davidson, who has said that "the only way tradition can be carried on is to keep inventing new things." It is in that spirit that printmaker and graphic artist Phyllis Burks has prepared the illustrations for this book.

NARRATIVES

This we know. The Earth does not belong to man; man belongs to the Earth. This we know. All things are connected like the blood which unites one family. All things are connected. Whatever befalls the Earth befalls the sons of the Earth. Man did not weave the web of life, he is merely a strand in it. Whatever he does to the web, he does to himself.

—Attributed to Chief Sealth (Seattle)

There is a search going on, some would call it a ricourso, *for the old themes—the old stories which are once again becoming the new themes of discovery. Interpreting texts and myths, attempting to understand ritual and ceremony, imbibing the nectar of rich oral histories, all these and still other expressions are finding their way into contemporary narratives on wildlands and wilderness.*

In this section you will find a variety of essays and narratives that portray connections between humans and place and speak to the palpable presence of the wild. Through narrative and direct experience of wildness we connect with the earth, not as resource, but as source. Learning the local vernacular is singing the songs of our own place, living within limits imposed by natural boundaries, and developing reciprocal relationships with other-than-human life.

These writings draw deeply on the contributors' personal faculties of observation and intuition. They are soundings, testing the depths of new/old waters, in territories beyond the reach of conventional "dashboard" knowledge. These explorations of place do not rely on traditional, anthropocentric values which place human needs at the top of the pyramid. Rather, through narrative lenses, the writers suggest that life's meaning and purpose flow through indigenous, unmediated relationships with the contents of home. For them, wildness is nature's authentic voice.

Years ago, Henry Beston wrote in his provocative book The Outermost House:

We need another and wiser and perhaps more mystical concept of animals. . . . We patronize them for their incompleteness, for their tragic fate of having taken form so far below ourselves. And therein we err, and greatly err. For the animal shall not be measured by man. In a world older and more complete than ours they moved finished and complete, gifted extensions of the senses we have lost or never attained, living by voices we shall never hear. They are not brethren, they are not underlings, they are other nations, caught with ourselves in the net of life and time, fellow prisoners of the splendor and travail of the Earth.

The Far Outside

GARY PAUL NABHAN

Any good poet, in our age at least, must begin with the scientific view of the world; and any scientist worth listening to must be something of a poet, must possess the ability to communicate to the rest of us his sense of love and wonder at what his work discovers.

EDWARD ABBEY, *The Journey Home*

I was in a small room in Alaska when I heard it. That was part of the trouble. I was supposed to be paying attention to what was being said in the room; after all, this was a nature writing symposium. But from where I sat I could hear ravens coming in to roost in the spruce trees above us, and wondered how their calls were different from those of the Chihuahuan ravens down where I live. I could look out the windows and see bald eagles swooping over the waters of the sound. Worse yet, I already had the stain and smell of salmonberries on my hands, and had been perplexed all morning as to why the ripe berries on two adjacent bushes were entirely different colors.

It was then that I heard it. A familiar warble came out of the well-educated, widely read humanist a few chairs away from me. She asserted a truism I had heard in one form or another for nearly thirty years:

"Each of us has to go *inside* before we can go *outside*! How can we give any meaning to the natural world until each individual finds out who he or she is as a human being, until each of us finds our own internal source of peace?"

Queasy, I immediately felt nauseous, indisposed. Something she said had stuck in my craw. Instantly, I was so out of sorts I had to leave the room. Our moderator followed me out to the porch, where I gasped for air.

"Are you *okay?*" she asked earnestly. "You looked *green* all of a sudden."

"I dunno." I breathed deeply and looked up at the crisp blue sky. "I must be . . . uh . . . under the weather a little. Let me see if some fresh air will help. . . . If you don't mind, I had better go for a walk."

As I ambled along, I wondered what had set me off. I wandered around on a rainforest trail, trying to spiral in on what in that room had disoriented me. First, I felt uncomfortable with the notion that we can give the natural world "its meaning." The plants and animals which I have observed most diligently over twenty years as a field biologist hardly seem to be waiting for me to give *them* meaning. Instead, most humans want to feel as though *we* are meaningful, and so we project *our* meanings upon the rest of the world. We read meaning into other species' behavior, but with few exceptions they are unlikely to do the same toward us.

Humans may, in fact, be rare even among primates in the attention we give to a wide range of other species' tracks, calls, and movements. To paraphrase one prominent primatologist: "If their inattention to their neighbors other than predators is any indication, most monkeys are extremely poor naturalists." The same can be said of many other wild animals which live in sight of, and in spite of, human habitations.

While it may somehow be good for *us* to think and write about plants and animals, I am reminded of John Daniel's humbling insight while hopping through a snake-laden boulder field: the snakes were not fazed by his thoughts, fears, or needs. As Daniel writes in *The Trail Home*: "The rattlesnakes beneath the boulders instructed me, in a way no book could have, that the natural world did not exist entirely for my comfort and pleasure; indeed, that it did not particularly care whether my small human life continued to exist at all."

Walking along, my restlessness increased as I considered the premise put forth in that room: the shortest road to wisdom and peace with the world is that which turns inward. I will not argue that meditation, psychotherapy, and philosophical reflection are unproductive, but I simply can't accept that inward is the only or best way for everyone to turn. The more disciplined practitioners of contemplative traditions can turn inward and still get beyond the self, but many others simply stumble into self-indulgence.

As Robinson Jeffers suggested over a half century ago, it may be just as valid to turn outward: "The whole human race spends too much emotion on itself. The happiest and freest man is the scientist investigating nature or the artist admiring it, the person who is interested in things that are not human. Or if he is interested in human beings, let him regard them objectively as a small part of the great music."

Finishing my walk among the great music of crashing waves and hermit thrushes, I conceded that the wisest, most inspired people I knew had all taken this second path, heading for what I call the Far Outside. It is the path found when one falls into "the naturalist's trance," the hunter's pursuit of wild game, the *curandera*'s search for hidden roots, the fisherman's casting of the net into the current, the water-witcher's trust of the forked willow branch, the rock climber's fixation on the slightest details of a cliff face. Oddly, it is hanging onto that cliff, beyond the reach of the safety net of civilization, where one may gain the deepest sense of what it is to be alive. As arctic writer and ethnographer Hugh Brody says of his predilection for working in the most remote human communities and wildest places he can find, "it is at the periphery that I can come to understand the central issues of living."

Unlike conditions within the metropolitan grid where it seems we have got nature surrounded, the Far Outside still offers the comic juxtapositions, the ones worthy of a Gary Larson cartoon. The flood suddenly looms large before Noah can get his family onto the ark full of animals; the bugs in the test tube have the last say about the entire experiment.

INWARD AND OUTWARD PATHS

When I returned home to the Stinkin Hot Desert, I had an urge to see how an elder from another culture might view this apparent dichotomy between inward and outward paths—or for that matter, the dichotomy between culture and nature. I drove a hundred miles across the desert to see a seventy-four-year-old O'odham farmer who had worked all his life "outdoors": tending native crops, chopping wood, driving teams of horses, gathering cactus fruit, hunting, and building ceremonial houses for his tribe's rain-bringing rites. He was consistently wise in ways that my brief bouts with Jungian analysis, *zazen* practice, and Franciscan prayer had not enabled me to be. And I knew that because he'd had a brush with death over the last year, he had been made sedentary and forced to be alone, and at home, for a longer time than ever before in his entire life. He sat outside on an old wooden bench, a crutch on either side of him, looking out at a small field which he would not be able to plant this year. I asked him what he had been working over in his mind the last couple months.

"I'd like to make a trip," he said nonchalantly for a man who had only traveled once beyond the limits of the desert—all the way to Gallup—and who now lived at the end of his life less than thirty miles from where he was born.

"Yes, before I die, I'd like to go over there to the ocean," he nodded to the southwest, where the Sea of Cortez lay a hundred miles away. It was a sacred place for the desert O'odham, where they used to go as pilgrims for salt and for songs. My elderly friend paused, then continued.

"Yes, I would like to hear the birds there in the sea. I would like to hear those ocean birds sing in my native language."

"In *O'odham ha-neoki?*" I asked. I must have looked surprised he felt the birds spoke in *his* language, for he then offered to explain his comment as if it had been scribbled in a shorthand indiscernible to me.

"Whenever my people used to walk over there to the ocean for salt, they would stand on the edge and listen to those birds sing. And they are in many of the songs we still sing today, even though we haven't walked or ridden horses there since the hoof-and-mouth quarantines in the forties. In the old days, they didn't start to sing those songs while they were still at the ocean. No, the people would go back home, and then some night, those ocean birds would begin singing in their dreams. That's where our songs come from. They would come to our medicine men, from the ocean, in their dreams. Maybe the ones who play the violin would hear them in their sleep, and their voices would turn up in their fiddle tunes. Maybe the *pascola* dancers would hear the way they flew, and it would end up in the way they sounded when they danced with their rattles. Those birds have ended up in our songs, and I want to hear them at the ocean before I die."

What struck me about my friend's last request was his desire to hear those birds for himself at the edge of the ocean. For a lifelong dweller in a riverless desert, the ocean must be a landscape wilder than the imagination, truly unfathomable. In the end, he sought to juxtapose his culture's aural imagery of ocean birds with what the birds themselves were saying. He desired to experience nature directly, as a measure of the cultural symbols and sounds he had carried with him most of his life.

My friend's songs and stories are conversant with and responsive to what we often refer to as "outer reality." This larger landscape is not superfluous or irrelevant to his culture's literature, music, or ways of healing. When I arrived at his home once, years ago, I saw him carrying into the kitchen a mockingbird which he had captured in a seed trap, killed, and carefully butchered, in order to cook the meat up and feed it to his grandson. Mockingbirds are not simply good mimics, they are irrepressibly loquacious; his grandson was not. In fact, the boy was nearly three years old and had not spoken a word. Concerned, my friend recalled the sympathetic ritual of his people for curing such difficulties: feed the mute one the songbird's flesh. He will have the best chance of being able to express himself if he ingests the wild world around him. In the O'odham language, the words for curing, wildness, and health come from the same root.

This is where "inner" and "outer" become not a duality but a dynamic—like every breath we take. We are *inspired* by what surrounds us; we take it into our bodies, and after some rumination we respond with *expression*. What we have inside us is, ultimately, always of the larger, wilder world. Nature is not just "out there," beyond the individual. The O'odham boy now has seed, bird, and O'odham history in his very muscles, in the cells of his tongue, in his reverberating voice box.

Lynn Margulis has recently pointed out that thousands of other such lives are literally inside each so-called human "individual." For every cell of our own genetic background that we embody, there are a thousand times more cells of other species within and on each of our bodies. It would be more fitting to imagine each human corpus as a diverse wildlife habitat than to persist with the illusion of the individual self. Or better, each of us is really a corpus of *stories*: bacteria duking it out for the final word in our mouths; fungi having clandestine affairs between our toes; other microbes collaborating to digest the world within our intestines; archetypal images from our evolutionary past roaming through our nerve synapses, testing out groin muscles against our brain tissue.

If I could distill what I have learned during a thousand and one nights working as a field biologist, waiting around campfires while mist-netting bats, running lines of live traps, or pressing plants, it would be this: each plant or animal has a story of some unique way of living in this world. By tracking their stories down to the finest detail, our own lives may somehow be informed, and perhaps enriched. The zoologist who radio-collars a mountain lion may call his research a range utilization analysis, but he is simply tracking that critter's odyssey. A botanist may refer to the adaptive strategy of a cactus, but only after carefully recording chapter and verse how the plant endures and prevails, despite droughts, freezes, or heat waves. An ecologist interested in the nutcracker's dispersal of pine seeds is slowly learning the language of the forest, and the birds are her newly found verbs.

Perhaps due to what Paul Ehrlich calls "physics envy," many biologists feel inclined to mask their recording of stories in shrouds of numbers, jargon, and theory. We find their remarkable insights buried beneath techno-babble about life histories, optimal foraging tests, or paleoecological reconstructions. Most of them, however, are merely tracing the trajectory of another life as it demonstrates ways to survive in the Far Outside. In *Writing Natural History*, two-time Pulitzer Prize winner E. O. Wilson tells of the struggle scientists have simply to be storytellers: "Scientists live and die by their ability to depart from the tribe and go out into an unknown terrain and bring back, like a carcass newly speared, some new discovery or new fact or theoretical insight and lay it in front of the tribe; and then they all gather and dance around it. Symposia are held in the National Academy of Sciences and prizes are given. There is fundamentally no difference from a Paleolithic campsite celebration. . . ."

In short, scientists too grapple with the challenge of telling the unheard-of stories which may move their tribes. And yet it is tragic to realize how few of these stories any of us will ever glimpse. In *The Diversity of Life*, it is E. O. Wilson again who reminds us that we have only the crudest of character sketches—let alone any understanding of the plots—involved with most of these floral and faunal narratives:

Even though some 1.4 million species of organisms have been discovered (in the minimal sense of having specimens collected and formal scientific names attached), the total number alive on the earth is somewhere between 10 and 100 million. . . . Of the species given scientific names, fewer than 10 percent have been studied at a level deeper than gross anatomy. [Intensively studied species make up] . . . a still smaller fraction, including colon bacteria, corn, fruit flies, Norway rats, rhesus monkeys, and human beings, altogether comprising no more than a hundred species.

STORIES

Try to imagine the still-untold stories, the sudden flowerings, the cataclysmic extinctions, the episodic turnovers in dominance, the failed attempts at mutualistic relationships, and the climaxes which took hundreds of years to achieve. In every biotic community, there are story lines which fiction writers would give their eyeteeth for: Desert tortoises with allegiances to place that have lasted upward of forty thousand years, dwarfing any dynasty in Yoknapatawpha County. Fidelities between hummingbird and montane penstemon that make the fidelities in Port William, Kentucky, seem like puppy love. Dormancies of lotus seeds that outdistance Rip Van Winkle's longest nap. Promiscuities between neighboring oak trees which would make even Nabokov and his Lolita blush. Or all-female lizard species with reproductive habits more radical than anything in lesbian literature.

And yet, with the myriad stories around and within us, how many of them do we recognize as touching our lives in any way? Most natural history essays are so limited in their range of plot, character development, and emotive currents that Joyce Carol Oates has come to an erroneous, near-fatal assumption about nature itself. In her essay "Against Nature," Oates claims that nature "inspires a painfully limited set of responses in 'nature writers' . . . *reverence, awe, piety, mystical oneness.*"

Most environmental journalists offer an even more limited set of "news" stories: (1) that someone has momentarily succeeded in disrupting the plans of the bastards who are ruining the world; and (2) that the bastards are still ruining the world. Most newspaper and magazine journalists who ostensibly cover biological diversity tell the same doom and gloom story over and over, with virtually nothing substantial about the nonhuman lives embedded in that diversity. One week, "Paradise Lost" is told with the yew tree as the victim in the temperate rainforest; the next, the scene has shifted to peyote in the Chihuahuan desert; but the plot is still the same.

I believe that human existence is being degraded by our ignorance of these diverse stories. In stark contrast to the O'odham elder's dreams, fewer and fewer creatures are inhabiting the dreams of those in mainstream society. I know another elderly man who lives in the midst of metropolitan Phoenix. Although he is a few years younger than my friend the Indian farmer, he seems far closer to death; I can feel it every time I visit him. He too was formerly an outdoorsman and farmer, skilled with horses, hunting, building, and wood carving. But now he has emphysema and cannot even go outside and sit, the contaminated air of Phoenix is so vile. Yet that is not all that is killing him. Confined to a hermetically sealed tract house, he sits in front a television all day long and hears just three stories repeated ad nauseam: (1) Saddam Hussein and other foreign despots are out to get us; (2) substance-abusing street gangs are out to get us; and (3) mutant microbes are out to get us. He seems drained of all resilience, a man without hope. He has lost all contact with the wildlife, the Far Outside, that had been his source of renewal most of his life.

Harking back to William Carlos Williams, we might say that society pays little attention to these myriad lives, but people die for lack of contact with them every day. As with our teeth, what we don't pay attention to is likely to disappear. By the end of this decade, twenty-five thousand species—twenty-five thousand distinctive ways of living in this world—are likely to be lost unless we begin to learn of these beings in ways that move us sufficiently to curtail our destructive habits.

And scientists cannot do the work by themselves. As E. O. Wilson admits, the capacity to tell of these vanishing lives in compelling ways is tightly constrained by the stylistic conventions of technical scientific journals. In *Writing Natural History* he argues that

the factual information that we get and the new metaphors created out of science somehow have to be translated into the language of the storyteller—by film, by speech, by literature, by any means that will make it meaningful and powerful for the human mind. . . . And the storyteller has always had this central role in societies: of translating that information in forms that played upon the great mythic themes and used the rhythms and the openings . . . the body . . . and the closures that make up literature.

Now, more urgently than ever before, we all need to come face to face with other lives in the Far Outside—not just with the Bali Mynah and the Furbish Lousewort, but with the fungi between our toes as well. Imagine what might happen if some of those who now turn inward, apprenticing themselves to all kinds of gurus, priests, therapists, and masters, would turn outward as apprentices to other species: Komodo Dragons, Marbled Murrelets, Desert Pupfish, Beer-Making Yeasts, Texas Wild Rice, or Okeechobee Gourds.

I can't help but wonder if the dilemma of our society is not unlike that of the mute child who needs to eat the songbird in order to speak. Unless we come to embody the songs from the Far Outside, we will be left dumb before an increasingly frightening world. But that is just the first step. Once we have begun to express in our own ways the stories inspired by those other lives, we may need to keep seeking them out, to constantly compare the images we have conjured up with the beings themselves.

It is time to leave this room and go Outside, farther than we have ever gone together before. It is time to hear the seabirds singing at the edge of the world and to bring them back, freshly, into our dreams.

Water Songs

TERRY TEMPEST WILLIAMS

Lee Milner and I stood in front of the diorama of the black-crowned night heron at the American Museum of Natural History in New York City. *Nycticorax nycticorax*: a long-legged bird common in freshwater marshes, swamps, and tidal flats, ranging from Canada to South America.

We each had our own stories. My tales were of night herons at the Bear River Migratory Bird Refuge in Utah—the way they fly with their heads sunken in line with their backs, their toes barely projecting beyond their tail, the way they roost in trees with their dark green feathered robes. Lee painted them at Pelham Bay Park on the northern edge of the Bronx, where, she says, "they fly about you like moths." Both of us could recreate their steady wingbeats with our hands as they move through crepuscular hours.

Two women, one from Utah and one from the Bronx, brought together by birds.

We were also colleagues at the American Museum. I was there as part of an exchange program from the Utah Museum of Natural History, on staff for six weeks. Lee was hired to manage the Alexander M. White Natural Science Center while the program director was on medical leave. The center is a special hands-on exhibit where children can learn about nature in New York City.

We worked together each day, teaching various school groups about the natural history in and around their neighborhoods. In between the toad, turtle, and salamander feedings, we found time to talk. Lee was passionate about her home. She would pull out maps of Pelham Bay Park and run her fingers over every slough, every clump of cattails and stretch of beach that was part of this ecosystem. She would gesture with her body the way the light shifts, exposing herons, bitterns, and owls. And she spoke with sadness about being misunderstood, how people outside the Bronx did not recognize the beauty.

I wasn't sure I did.

Lee and her father had just moved to Co-op City, and she described the view from their apartment as perfect for looking out over cattails. She promised to take me to Pelham Bay before I left.

The opportunity finally came. Our aquarium had been having bacteria problems that had killed some of the organisms. We decided we could use some more intertidal creatures: crabs, shrimps, and maybe some barnacles. We needed to go collecting. Pelham Bay was the place.

David Spencer, another instructor, agreed to come along. The plan was to meet Lee at Co-op City in the morning. David and I packed our pails, nets, and collecting gear before leaving the museum for the bus. Our directions were simple—one crosstown transfer, a few blocks up, and we were on the Fordham Road bus to Co-op City.

The idea of finding anything natural in the built environment passing my window seemed unnatural. All I could see was building after building, and beyond that, mere shells of buildings burnt out and vacant with empty lots mirroring the human deprivation.

"South Bronx," remarked David as he looked out his window.

Two elderly women sitting across from us, wrapped in oversized wool coats, their knees slightly apart, smiled at me. I looked down at my rubber boots with my old khakis tucked inside, my binoculars around my neck and the large net I was holding in the aisle—how odd I must look. I was about to explain when their eyes returned to their hands folded neatly across their laps. I asked David, who was reading, if he felt the slightest bit silly or self-conscious.

"No," he said. "Nothing surprises New Yorkers." He returned to his book.

We arrived at Co-op City. Lee was there to meet us. I was not prepared for the isolating presence of these high-rise complexes that seemed to grow out of the wetlands. Any notion of community would have to be vertical.

From her apartment, Lee had a splendid view of the marshes.

Through the haze, I recognized the Empire State Building and the twin towers of the World Trade Center. The juxtaposition of concrete and wetlands was unsettling. They did nothing to inspire each other.

"The water songs of the red-winged blackbirds are what keep me here," Lee said as we walked toward Pelham Bay. "I listen to them each morning before I take the train into the city. These open lands hold my sanity."

"Do other tenants of Co-op City look at the marsh this way?" David asked.

"Most of them don't see the marsh at all," she replied.

I was trying hard not to let the pristine marshes I knew back home interfere with what was before us. The cattails were tattered and limp. Water stained with oil swirled around the stalks. It smelled of sewage. Our wetlands are becoming urban wastelands. This one, at least, had not completely been dredged, drained, or filled.

It was midwinter, an overcast sky. The mood was sinister. But I trusted Lee, and the deeper we entered into Pelham Bay Park, the more hauntingly beautiful it became, in spite of the long shadows and thin silhouettes of men behind bushes.

"This is a good place for us to collect," she said, putting down her bucket at the estuary.

Within minutes, we were knee-deep in tide pools and sloughs. My work was hampered by the muck that leached into the water. I could not see, much less find, what one would naturally assume to be there. More oil slicks. Iridescent water. Yellow foam. I kept coming up with gnarled oysters with abnormal growths on their shells. I handed an oyster dripping with black ooze to David.

"Eat this," I said.

"Not until you lick off your fingers first," he replied, wiping the animal clean.

These wetlands did not sparkle and sing. They were moribund.

Lynn didn't see them this way. She knew too much to be defensive, yet recognized her place as their defender and saw the beauty inherent in marshes as systems of regeneration. She walked toward us with a bucket of killifish, some hermit crabs, and one ghost shrimp.

"Did you see the night heron?" she asked.

I had not seen anything but my own fears fly by with a few gulls.
We followed her through a thicket of hardwoods to another clearing.
She motioned us down in the grasses.

"See him?" she whispered.

On the edge of the rushes stood the black-crowned night heron. Perfectly still. His long white plumes, like the misplaced hairs of an old man, hung down from the back of his head, undulating in the breeze. We could see his red eye reflected in the slow, rippling water.

Lee Milner's gaze through her apartment window out over the cattails was not unlike the heron's. It will be this stalwartness in the face of terror that offers wetlands their only hope. When she motioned us down in the grasses to observe the black-crowned night heron still fishing at dusk, she was showing us the implacable focus of those who dwell there.

This is our first clue to residency.

Somehow, I felt more at home. Seeing the heron oriented me. I relaxed. We watched the mysterious bird until he finally outpatienced us. We left to collect a few more organisms before dark.

I made a slight detour. I wanted to walk on the beach during sunset. There was no one around. The beach was desolate, with the exception of a pavilion. It stood on the sand like a forgotten fortress. Graffiti looking more like Japanese characters than profanities streaked the walls. The windows, without glass, appeared as holes in a decaying edifice. In the middle of the promenade was a beautiful mosaic sundial. Someone had cared about this place.

In spite of the cold, I took off my boots and stockings and rolled up the cuffs of my pants. I needed to feel the sand and the surf beneath my feet. The setting sun looked like the tip of a burning cigarette through the fog. Up ahead, a black body lay stiff on the beach. It was a Labrador. Small waves rocked the dead dog back and forth. I turned away.

Lee and David were sitting on the pavilion stairs watching more night herons crisscross the sky. Darkness was settling in. Lee surmised we had wandered a good six miles or so inside the park. Even she did not think we should walk back to Co-op City after sunset. They had found a phone booth while I was out walking and had called a cab.

"So are we being picked up here?" I asked.

They looked at each other and shook their heads.

"We have a problem," David said. "No one will come get us."

"What do you mean?"

"The first company we dialed thought we were a prank call," said Lee. "Sure, you're out at Pelham Bay. Sure, ya'll want a ride into the city. No cabby in hell's dumb enough to fall for that one . . . click."

"And the second company hung up on us," David said. "At least the third cab operation offered us an alternative. They said our only real option was to call for a registered car."

"Let's do it," I said.

"We would have except we've run out of change," Lee replied.

I handed her what I had in my pockets. She called a gypsy cab service.

Waiting for our hired car's headlights to appear inside this dark urban wilderness was the longest thirty minutes I can remember. We stood on the concrete steps of the pavilion like statues, no one saying a word. I thought to myself, "we could be in Greece, we could be in a movie, we could be dead."

The registered car slowly pulled up and stopped. The driver pushed the passenger's door open with his foot. Because of all of our gear, I sat in front. Our driver could barely focus on our faces, let alone speak. I noticed his arms ravaged with needle tracks, how his entire body shook.

Eight silent miles. Thirty dollars. I gave him a generous tip and later felt guilty, knowing where the money would go. David and I hugged Lee, thanked her, and took the specimens, buckets, screens, and nets with us as we caught the bus back to Manhattan. The hour-long ride back to the city allowed me to settle into my fatigue. I dreamed of the pavilion, the stiff black dog, and the long-legged birds who live there.

David tapped me on the shoulder. I awoke startled. Disoriented.

"Next stop is ours," he said.

We got off the bus and walked over to Madison and Seventy-ninth Street to catch the crosstown bus back to the museum. We kept checking the fish to see if they were safe, surprised to see them surviving at all given the amount of sloshing that had occurred throughout the day.

As we stood on the corner waiting, a woman stopped. "Excuse me," she said. "I like your look. Do you mind me asking where you purchased your trousers and boots? And the binoculars are a fabulous accessory."

I looked at David who was grinning.

"Utah," I said in a tired voice. "I bought them all in Utah."

"I see . . ." the woman replied. "I don't know that shop." She quickly disappeared into a gourmet deli.

Back at the museum, the killifish were transferred safely into the aquarium with the shrimp and crabs. Before we left, I placed the oysters in their own tank for observation. With our faces to the glass, we watched the aquariums for a few minutes to make certain all was in order. Life appeared fluid. We turned off the lights and left. In the hallway, we heard music. Cocktail chatter. It was a fund-raising gala in the African Hall. We quietly slipped out. No one saw us enter or leave.

Walking home on Seventy-seventh Street, I became melancholy. I wasn't sure why. Usually after a day in the field I am exhilarated. I kept thinking about Lee, who responds to Pelham Bay Park as a lover, who rejects this open space as a wicked edge for undesirables, a dumping ground for toxins or occasional bodies. Pelham Bay is her home, the landscape she naturally comprehends, a sanctuary she holds inside her unguarded heart. And suddenly, the water song of the red-winged blackbirds returned to me, the songs that keep her attentive in a city that has little memory of wildness.

Biodiversity, Wildness, and The Wildlands Project

REED F. NOSS

I grew up spending every available moment outdoors—catching snakes, turtles, and frogs and looking them up in my Golden Guides; scooping up minnows in strainers borrowed from Mom's kitchen; risking life and limb swinging on grapevines over steep ravines; exploring limestone caves and collecting fossils in an endless search for the perfect trilobite. Unlike other boys, baseball and television held little attraction for me. I didn't like fishing because it made me feel sad to take a life or even to see a hook in a fish's mouth. I was both a daredevil and a wimp. My friends were those who were willing to take a break from baseball or television and spend a few hours exploring the wild. Most of these friends were temporary. My real companions were the trees, the flowing waters, all kinds of creatures in the woods and streams, and imaginary creatures in the clouds. None of them ever let me down.

There, in the suburban-rural fringe of southwestern Ohio in the 1950s and early '60s, nature was always mysterious, wonderful, and personal. It was also, I soon realized, threatened. I remember one perfect day in late summer. A friend and I were netting darters and other small fish in Hole's Creek, when we noticed a slick, shiny film on the surface of the water. Then we began seeing dead fish—the darters, minnows, and shiners we had been catching and releasing were floating in an oily film. As we waded back up the creek later that day we discovered a place where a landowner had cleaned out his garage and thrown all his old cans of paint, turpentine, oil, and pesticides into the creek. Our creek had been violated and we were furious. On the way home on our bikes we made a vow to get even. The next day we threw rocks through the windows of the landowner's garage, then ran like hell down the creek.

Of course, it was impossible to get even. The damage had been done. The landowner probably never suspected that the rocks which broke his windows were thrown in revenge for his mistreatment of a living stream. Other assaults on our woods were more pervasive and quickly became unstoppable. Bulldozers began clearing home sites, knocking down the great trees we climbed and filling the streams with sediment. On several occasions my friends and I came upon idled bulldozers in the woods and spontaneously we smashed large rocks against them and ripped out whatever hoses and wires we could grab from the engines. None of this probably had much effect on construction schedules. It didn't even make us feel all that much better. But at least we tried.

Were these senseless acts of youthful vandalism? Or were they ethically legitimate attempts, however immature and futile, to defend what we loved? I cannot answer such questions objectively, but I will say my conscience is not the slightest bit guilty. Obviously we didn't stop the bulldozers. The woods we loved and played in have been displaced by large, green chem-lawns with ostentatious houses. All "brush" and other messiness of the natural system has been removed. There are no more snapping turtles or rainbow darters in Hole's Creek. You would be lucky to find a creek chub or a water strider. Children growing up in that neighborhood today don't have the opportunities to enjoy nature that I had. Presumably they watch television instead. After all, they can see creatures more exotic than snapping turtles on TV nature shows.

Not all of my experiences of nature as a child ended in damage or melancholy. In fact, my memories are largely ones of joy—joy in finding a new kind of snake or salamander, in turning over that creek rock and discovering a near-perfect trilobite fossil, and in simply experiencing the beauty and wildness of a beech-maple forest. As I grew older and even less concerned about being thought a sissy, I was amazed to discover the diversity of wildflowers, butterflies, and songbirds that I had previously neglected. By the time I was in college, I could still find places that were wild and reasonably intact biologically, but I had to wander farther from home. Sadly we have to wander farther and farther to experience wild nature, unable to stay in one place, put our feet down, and say to the developers or loggers, "Stop! You've taken enough." Bioregionalism is not easy.

BIODIVERSITY

My experiences as a naturalist and simply as a kid playing in the woods have shown me that there are two things I value most in nature: diversity and wildness. Biological diversity, or biodiversity as we now call it, is the basis of all natural history and ecological science. If there were only a couple kinds of animals and plants out there, I doubt anyone would want to be a naturalist. The allure is the seemingly endless variety of forms and processes. Now that this diversity is threatened and many species are becoming rare, we have developed conservation biology, the "science of scarcity and diversity" as Michael Soulé describes it in a book title. Ecologists and conservationists have tried to justify their interest in biodiversity by claiming that diverse ecosystems are more stable and by suggesting that diversity has utilitarian value—who knows which species holds within its chemistry the cure for cancer? But as David Ehrenfeld points out, utilitarian arguments are not entirely satisfactory and are often quite unconvincing. "What biologist," Ehrenfeld asks, "is willing to find a value—conventional or ecological—for all 600,000-plus species of beetles?"

Why not admit that we value biodiversity for its own sake? Why not state openly what is intuitively obvious: each species has inherent worth? Thinking back as far as I can, I never remember believing anything different. Now, as a scientist, thinking otherwise still makes no rational sense to me. Where is the objective support for the widespread belief that humans are fundamentally superior to other beings? Why is a large brain with a convoluted cortex any more a sign of superiority than the ability to dive like a peregrine falcon at 200 miles per hour, construct huge living reefs like the corals, or live symbiotically with the roots of great trees like mycorrhizal fungi? Each of these evolutionary strategies has proved successful, yet paradoxically ours may prove self-defeating as we destroy the Earth and ourselves with it. The only thing that might stop us from devastation, I believe, is the ethical and rational appreciation that each living thing is just as important as we are. It follows that we have an ethical obligation to protect our evolutionary kin when they are imperiled by our civilization.

WILDNESS

But what about wildness? Scientists have accepted the need to preserve biodiversity and have written thousands of papers about how we might go about doing it. We have left wildness to the poets and mystics. But isn't there a poet and mystic inside each of us? What might the right side of our brain tell us about wildness if we took the time to listen? Envision a tame world in which all biodiversity is intact within a system of zoos, botanical gardens, intensively managed forests and rangelands, and scattered parks and preserves. Such a world would be very safe. We would still have grizzly bears, wolves, tigers, and other dangerous animals, but they would be securely confined in cages. Insect pests would be kept under tight control. Human disease organisms would be bottled. Our parks would have paved trails with guard rails near dangerous cliffs or chasms. We could experience nature without ever having to worry about getting lost, freezing to death, getting mauled by a bear, being besieged by mosquitoes, or even stepping in mud.

I don't believe such a thoroughly tamed world is possible. Nature will always be capable of surprising us. I am immensely thankful for earthquakes, volcanoes, hurricanes, floods, wildfires, and new strains of viruses. But it is certainly reasonable to conclude that nature is becoming tamer and safer for us in middle-class America all the time. Our opportunities to be humbled—put in our place—are diminishing rapidly. Indeed, I believe that wildness is being destroyed even faster than biodiversity and that the consequences for the human spirit are even more grim. As we eliminate opportunities to be lost, mauled, frozen, stampeded, infested, or otherwise dethroned, we become all the more arrogant, all the more certain of our ability to control nature, to exploit those species we find economically useful or aesthetically appealing and get rid of all the rest—the worthless. Thus the ultimate consequence of losing wildness will be a world severely impoverished.

Yet wildness by itself is not a satisfactory criterion for conservation.

Most of the remaining wild areas are places with severe environmental conditions (steep slopes, infertile or erodible soils, harsh weather) making them uncomfortable or unprofitable to develop. Although these wild areas tend to be low in biodiversity, they are often important as refugia for carnivores and other species exploited or persecuted by humans. When wild areas are tamed, these species are lost. Thus biodiversity and wildness are truly complementary: they need each other and we need them both.

THE WILDLANDS PROJECT

I believe we have a moral imperative to try to restore biodiversity and wildness across the globe, however utopian this venture might seem. This is why I helped found The Wildlands Project. Current efforts to preserve biodiversity are not maintaining wildness. Nor are they adequate to maintain the full richness of genetic material, species, habitats, ecosystems, and ecological and evolutionary processes. Conventional conservation efforts, which established medium-sized to fairly large reserves to protect scenery and recreational opportunities and, most often, only small reserves for biological or scientific purposes, have proved inadequate. The new whiz-bangs—"sustainable development," "extractive reserves," and "ecosystem management" (as interpreted by most agencies and politicians)—are every bit as deficient because they destroy wildness and in the process remove species that are highly vulnerable to human disturbances.

The Wildlands Project has four major goals for each region: representing all types of ecosystems across their natural range of variation in protected areas; maintaining viable populations of all native species, with most attention to large carnivores and other species especially sensitive to human activities; sustaining the full suite of ecological and evolutionary processes; and creating a conservation system that is adaptable to a changing environment. I have described these goals in detail elsewhere (see the References) and only wish to point out here that they are generally accepted among conservationists. Less well accepted (or even talked about) are the implications of these goals—for example, how much land must be protected to provide for species viability?

Laws such as the Endangered Species Act and the National Forest

Management Act are ostensibly designed to maintain viable populations and keep species—all species—from going extinct. But government agencies do not seem to take these goals very seriously. A recent analysis of U.S. Fish and Wildlife Service recovery plans published in *Science* magazine showed that 28 percent of recovery goals for endangered and threatened species were set at or below population sizes at the time the plans were written; 60 percent of vertebrates have recovery goals below what biologists Georgina Mace and Russ Lande have recently recommended as the threshold for endangered status. Every working conservationist knows that the U.S. Fish and Wildlife Service rarely enforces the "taking provision" (Section 9) of the Endangered Species Act, especially on private lands. (Section 9 is supposed to prevent habitat destruction as well as direct killing of listed species.) Why are our government agencies behaving negligently? Because doing what it really takes to restore viable populations of endangered species would be bad for business. We would have to put one hell of a lot of land "off limits" to developers.

The Wildlands Project insists that we take our obligations to restore populations and ecosystems seriously. We must determine what is really needed—in terms of land area protected and other changes regarding how human society interacts with nature—to achieve these goals. We want to put it all on the table. If it takes 32 million acres of wilderness in the northern Rocky Mountains to maintain a viable population of grizzly bears, which recent analyses suggest is probable, then let's not hide from that politically sensitive fact. Let us confront the implications of our science and our ethics head-on. I've heard rumors that The Wildlands Project, because it endorses a forthright approach, is giving conservation biology a bad name. I'm saddened by these rumors. But I suspect that those who worry about conservation biology being seen as unreasonable are the same people who believe that science is value-free and that scientists should never address the policy implications of their studies. I heard that kind of nonsense when I worked for the EPA. And I left. I believe that scientists can be honest, objective, rigorous, and still work with a sense of mission. That mission for me is restoring wildness and biodiversity.

What is so perplexing to some people about The Wildlands Project is that we have no hidden agenda. Special interests often accuse environmentalists of wanting something more than they ask for. "You don't just want to reform forestry," they say, "you want to lock up half the country as wilderness." The Wildlands Project would reply: if that is what's needed to restore the natural heritage of our country, then that is precisely what we want. Current estimates of what it takes to represent all species and ecosystem types in protected areas and maintain viable populations of large carnivores generally range on the order of 25 to 75 percent of a given region and sometimes higher. By way of comparison, less than 5 percent of the United States and the world is currently protected in strict reserves. The protection targets set by the IUCN and the Brundtland Commission are 10 and 12 percent, respectively. These low estimates have no scientific basis. They are apparently motivated by little else than a desire to appear politically reasonable. I am troubled that even most conservation biologists are afraid to state publicly what is needed, according to the best available science, to protect biodiversity.

Concluding that 25 to 75 percent of a region must be protected in order to meet well-established conservation goals is not the same as saying that this much land must be "locked up" within designated parks or wilderness areas. It is at least conceivable that we can manage land for multiple uses and still maintain most elements of biodiversity. But the record of multiple-use management so far is bleak. The most sensitive components of the biota—for example, grizzly bears and wolves—generally persist only within landscapes with very low road density and minimal human development. Often this means wilderness—defined as roadlessness and absence of permanent human habitation. Obviously many regions have already lost the large carnivores and some have less than 5 percent of the land remaining in a natural state. For these regions, less ambitious goals must suffice for the short term. It may be several hundred years before we have wolves back in Ohio. But I can picture them now, chasing down elk in the abandoned suburbs, denning in the crumbling yuppie mansions, the pups playing in the creek where I found my first trilobite, who lived there 450 million years ago.

When we take the long-term view, encouraged by ecology and

evolutionary biology, then recovery of wilderness and wildland-dependent species almost anywhere becomes a reasonable goal. The human population cannot grow forever; it must soon stabilize. Looking far enough ahead, human population decline is probable; therefore, ecological recovery can become a reality. What we are trying to do now is to save as much biodiversity as we can so that recovery can be more rapid and complete. The glimmer of hope that the scale of human influence will decline and we will reestablish a harmonious relationship with the Earth is the foundation of The Wildlands Project. However unlikely an ecotopian future may seem, what is the alternative? Last-ditch efforts to save a few remnant natural areas? Dignified retreat? Despair? Despair is not much of a motivator. Hope for the future is what keeps me working as a conservationist.

KNOWING THE LIMITS

I have told you a little about how I came to appreciate wildness and biodiversity. These two concepts are not interchangeable, but each for its full expression requires the other. We can have diverse landscapes that are well settled by people. These landscapes may contain all but a few of the native species present before the European invasion. I would not be upset if these landscapes were a substantial part of our future milieu. Aesthetically, I often enjoy the agrarian, semiwild landscapes that René Dubos and others have described as the ideal future condition. Even cows and plows have their place. But we have far too many of them. As Thoreau suggested, we need to witness our own limits transgressed. The Wildlands Project insists that limits on civilization be reconstructed in each region on Earth. Grizzly bears, wolves, jaguars, and crocodiles need these limits.

If we fail to impose the limits that a "land ethic" (in the sense of Aldo Leopold) requires, the world will go on. With or without people, life on Earth will likely persist until the sun expands into a red giant, or until a huge comet intercepts our orbit. The final impact of humans on the biota might be nothing more than another downward blip in the fossil record, just one of many mass extinctions that will be well compensated for by another 20 or 30 million years of evolution. No one will miss us or curse us in the millions of years after our species' demise. But wouldn't it be better to go out knowing that we showed some respect for our fellow creatures while we were here? Despite the utter indifference of nature to all we have done, isn't loving her the right thing to do?

References

Ehrenfeld, D. "Why Put a Value on Biodiversity?" In E. O. Wilson, ed. *Biodiversity.* Washington: National Academy Press, 1988.
Leopold, A. *A Sand County Almanac.* New York: Oxford University Press, 1949.
Noss, R. "A Regional Landscape Approach to Maintain Diversity." *Bioscience* 33 (1983):700–706.
———. "Wilderness Recovery: Thinking Big in Restoration Ecology." *Environmental Professional* 13 (1991):225–234.
———. "Biologists, Biophiles and Warriors." *Wild Earth* (Winter 1991/92):56–60.
———. "The Wildlands Project: Land Conservation Strategy." *Wild Earth* (Special Issue 1992):10–25.

The Wild and the Tame

STEPHANIE MILLS

Late in August 1993, in the throes of writing, I had the strangest dream: I had to cut in horizontal thirds a copy of my first book, whose title asked, but did not answer, the question *Whatever Happened to Ecology?* I applied a breadknife to the task. When I opened the three pieces, I discovered the bottom third consisted of the blank space below the type. Then this fraction became a whole book, looking very much like Peterson and McKenny's *Field Guide to Wildflowers*.

You can't *not* think about a dream like that. What could it mean? The following evening I wandered out in the field with my scissors and my cat, wanting a free bouquet. But the gaillardia I planted on the slope were all done with their flaming ember show. Lemon beebalm, a goldenrod, blackeyed susans, and pearly everlastings were the only flowers remaining, and none of them appealed to the florist in me. Spotted knapweed, being abundant, but *planta non grata*, got a pass too. I did, however, find a russet and walnut striped feather, tipped dark, then white, that looked like it might recently have belonged to a hawk; and saw that a likely fox den had some fresh excavations. Sign of fox, sign of hawk, traces of the wild.

A possible interpretation of the dream occurred to me. To explain it requires me to take two leaps backwards in time: first, to 1970, when I titled a never-to-be-finished book I was working on, *Whatever Happened to Ecology?* At that time I thought the environmental movement was forsaking mystery for policy, wildness for respectability—selling out, in short. Hence the confrontational tone of the question. In 1985, when I took up work on the to-be-finished book of the same title, I had a less patent answer, wound up writing a memoir, something of an apologia for pat answers, a diffuse narrative of my personal journey into the bioregional movement, which is decentralist, place-located, ecosystem-based. Part of the answer, then, was that ecology and social change had become, for bioregionalists, mutually informing if not inseparable.

Now leap forward to 11 September 1992. It was my birthday, the day of a full moon. I was in little Tibet—Ladakh, now the northernmost part of Jammu and Kashmir State in India. A conference called "Rethinking Progress" took me there, to Leh, the capital city. Given some free time, another conferee and I took off for a village outside of Leh to consult Amalamu, a prospering oracle and healer. The séance was held in her kitchen, there was an apprentice with her, and although the way they went about inducing their trances was workmanlike, the shuddering, yipping, and convulsing cannot have been easy work.

My question to Amalamu was as convoluted as only an introvert's can be, having to do with whether the nature of the service I am called to do is artistic or political. It proved to be untranslatable in detail and got boiled down to: What should I do? What *do* you do, she wanted to know. I write about nature, I said. Keep it up, she said. "If you write a book about past, present, future of nature, your future will be bright."

Naturally it was nice to hear about the possibility of a bright future— standard soothsayer talk, I thought. But she had no obvious reason to advise me to think about the *future* of nature. Perhaps the top two-thirds of whatever happened to ecology—its past and present—are already set in type? They cannot be revised or rewritten. But the future of nature—the last third—is marginal, open, blank at first, then becomes an entire book, a guide to the wildflowers, a field guide. Perhaps even a "feeled" guide.

Like it or not, we are entering a new era in earth's history where wild nature—*autochthonous* (a mouthful of a word meaning indigenous from the Greek *autos*, self, and *chthon*, earth) being—may not persist at the macroscopic level except by human sufferance. In a valedictory address to the Society for Conservation Biology, Michael Soulé described the prospect with some irony:

Global warming, drying, ozone depletion, and toxification will produce many unpredictable ecosystem-level effects that will have dramatic impacts on biodiversity. Fragmentation will continue to harry habitats; its bitter fruits will provide many opportunities to discover and treat new kinds and degrees of area and edge effects and related maladies. The truncation of biological communities by the removal of top carnivores, important herbivores, plants that provide critical resources, and habitat-reforming taxa (beavers, elephants, pocket gophers, termites, etc.) will yield valuable information about the roles of keystone species. Restoration ecology and conservation biology will tend to merge because most so-called wild places on the planet will be relatively denatured and will need intensive rehabilitation and management.

I came into ecological concern at a time when valuing wild over tame was the rock upon which conservationists built their church. Thoreau's saying "in Wildness is the Preservation of the World," was a basic article of faith. I became, minus the hikes, a devotee of that holy wildness, understanding it to be the ground of all earthly being, including the human. Yet one of the most startling realizations of the present moment is that the living tissue of wildness—biodiversity—is utterly dependent on our care. The psychological and moral shock of assuming our existential responsibility for the preservation of species is terrific. To me it seems as though the ground of mind, god, and being are all shifting. For humans to locate themselves in right humble relation to wild nature under these changed circumstances, at the end of an era whose metaphors derived from mechanics and promoted the illusion of control, will require a radical rededication to the wild. Conservation biologist Reed Noss has observed that our species is more adaptable (and more destructive) than any other. Our blessing and curse of adaptability must now be put in service of restoration for the whole of nature, the health of nature, which is to say, wildness.

The "our" of *our* care currently includes 5.4 billion human beings. Wilderness buffs may number in the thousands; ecologically concerned individuals, ecosystem peoples, and members of traditional subsistence cultures number in the millions. Which means we have a whole lot of information exchange and consensus building—call it reinhabitory work—before us if we are to secure the future of nature. Every kind of effort has to go on simultaneously: solitary artisan work creating prairies, the polyglot expression of a land ethic, hard-core struggles to protect what remains of old-growth anything, be it forests or coral reefs.

The future of nature will be a function (albeit not exclusively) of the soundness of human community. We might prefer it otherwise but must concede that Homo Sappy's got the hole card now. And our species, so terribly alienated from the wild, so displaced from any community, is in as urgent need of healing as the land.

———

In the summer of '93, unusually heavy rains deluged the upper Mississippi River Valley. A so-called "natural" disaster—a devastating inundation of agricultural lands and cities built on lands claimed from the river's floodplain—wrought havoc on thousands of human lives. The economic effects of this disruption on concentrations of large-scale agricultural production, and the necessity to provide at least some government relief for individuals and communities whose lives were nearly destroyed, are effects which will likely be rippling through the system for years to come.

Understandably a lot of public reaction blamed the river, damned nature, and spoiled for another fight with the vast region's hydrology. There was some mention of the inevitability of flooding in floodplains. There was little mention of the deforestation of the watershed's uplands, the draining of its lowlands, the acres of pavement and almost impermeable lawn where once there had been deep-rooted prairie ecosystems. Compaction of the soil by farm equipment had, in truth, helped to create the disaster. These floods were the consequence of an aggregate of what is deemed "necessary" human activity, all undertaken with a degree of ecological obliviousness, if not plain ignorance. Even so, there were people throughout the region who loved living by whichever river so much that they planned to stay put or return as soon as they could, with cussed affection for their lately-wild watercourse.

Interestingly, a few voices were suggesting that since the Army Corps of Engineers had not succeeded in getting the river to behave rationally, it might be time to negotiate and allow some wetlands and floodplains to revert from human purposes to perform their timeless function of buffering the periodic changes in streamflows. Thinking like this might even portend a boom in restoration activity (as might the breakdown of so many other outsize attempts to redirect natural systems for exclusively human purposes, be they military, economic, or agricultural).

Even when attempts at domesticating, or subjugating, wild systems or organisms appear to have succeeded, the results are a little perverse. (Consider the passive gluttony of hatchery salmon, the flavorlessness of store-bought, midwinter tomatoes, and the foot-wetting neurosis of miniature poodles.) In contrast to these anomalies are the qualities of wildlife and wild places: authenticity, indigeneity, specificity, and spontaneity; resilience and health. Aldo Leopold spoke of land health—a quality he deemed rare as long as a half-century ago. Land health, "its perfect norm," is what we behold in wilderness, perhaps what we go to wilderness to behold.

––––––

There are two threads here: wilderness and reinhabitation. Wilderness, Americans who enjoy such surroundings have come to think, is rocks and ice, high remote challenging places. These gorgeous, albeit somewhat austere, leavings of land were what could be acquired for parks and wildlife refuges. But wilderness—wildness—is less a matter of topography than biological diversity and abundance. Wilderness, which until Columbus was what the New World was, is fat with life, with lives, both in numbers and kinds.

Wilderness, Nancy Newhall wrote, holds the answers to questions we have not yet learned to ask. In the train of so-called "natural" disasters (which really are failures of human programs to be resilient within the planet's incessant changing), those questions become survival questions, and not merely for the survival of our own species. Culture historian Thomas Berry says that if human consciousness had evolved on the moon, it would be as barren as the moon. As the ongoing industrial crusade to turn all earthly life to commercial purpose steadily impoverishes the biosphere and human cultures, our living images of graceful possibility dwindle.

What little wilderness remains displays the patterns to be restored—patterns for us to follow if our species (and as many others that still remain) are to persist here a while longer. Ideally this would call for a broad cultural reorientation to the wild, a long-overdue armistice in civilization's war upon it. Reinhabitation means learning the whole history of one's bioregion or watershed and developing a vision of sustainable ecological community out of that knowledge, as well as a vision of what we have been learning, in the last third of the century, about fine-tuning old-style household and neighborhood frugality. In addition to ending the hostilities to wilderness by reining in our natural wants and needs, there must be a surrender of conquered territory. Conservation biologists now posit that for there to be sufficient territory to allow the survival of North America's ecosystems and the species they comprise, about 50 percent of this land would have to be removed from human use and restored to a wilderness condition.

Substantial wilderness reserves are the heart and soul of the bioregion, epitomizing its character, existing as its sacred ground. Concentric zones of increasing human use and settlement density would embrace the ecosystem reserve. Livelihood would be patterned on and derived from the stability and economy of the climax community. This posits a postindustrial human culture and a postmodern indigeneity.

I don't doubt that the valiant engineers and hardworking farmers of the Upper Mississippi Basin knew some of the history of their region, possibly even of "agricultural" regions around the world that were finally bested by rivers and rebuked by exhausted soils. Probably the engineers felt they had no alternative but to try again, and possibly they felt that our remarkable postwar technologies would enable them to get it right this time. Somehow the possibilities to cooperate with great natural systems, rather than conquer them, seem much richer and more widely practicable. Reinhabiting bioregions would not be a function of a managerial elite, but would proceed in a manner that takes into account that the traditional lifeways of land-based tribal peoples worked well for millennia and left only modest traces in the land.

The hope of ecological restoration has begun to produce a subtle earth-change in my sense of the future. It has to do with our consciousness, with developing an informed and deliberate willingness to proceed into a millennium in which our long-accustomed relationships to nature are reshaped and in which we are reshaped by our dealings with nature. One still hopes for a liberatory future, a far gentler, indeed a happier, future. I believe that restoration will help us there. It is, I admit, a mystical faith in the epiphanic value of ecosystems.

In Wildness is the Preservation of the World: this insight is the why of it all. We cannot learn to be free within the confines of our own species. We cannot understand life and death and what they are for in exclusively human terms. Without that which is wild, the world becomes a cell block.

There is a quality of wildness in every living organism, in every great planetary process. Weather, tectonics, cell division, instinct—these are incalculable dynamics, phenomena of grace. My watershed-organizing friend Freeman House remarked as we stood on the beach watching the Mattole River enter the Pacific, "The air is wild; your intestines are wild." Life and its emergent properties are wild. The soul, perhaps, is wild. There is a sense in which certain kinds of human gatherings—affinity groups, study circles, watershed councils, even committees working by consensus—evolve by processes, ideas, and relationships that are novel, transcendent, surprisingly greater, more compassionate and livelier, than the sum of each individual's imaginings. Some essence of wildness will persist in us, unto our last breath. The microbes that return us to our elements are wild, too.

Irreducible essence of wildness aside, the duty of ecological restoration, as with conservation, is to ensure that wildness will be multifarious and at all scales, including the continental, that there will be a wild variety of habitats and species, and that the wild mystery of evolution may proceed. The struggle to allow this is now in its decisive throes.

In addition to admiring those who know the plants and grow the trees, I have utmost respect for wilderness defenders and traditional indigenous peoples striving to preserve their homelands. They are a dedicated and relentless kind. If it were not for their determination to fight for the sanctity of unspoiled places, and to keep on fighting despite having pieces of their hearts torn out with each loss, we would have no way of knowing wildness writ large and whole. There would be no refuges for beings larger than ourselves, no inedible or natural landscapes, no expanses great enough to shelter the shy and finicky creatures of the deep interior. Human concerns, important as they are, pale into repetitive melodrama alongside this epochal necessity to relinquish our pursuit of dominion over all of the earth. If as a species we can consciously evolve enough to preserve and restore wilderness, then wildness will secure for us a world.

———

Civilized humans have a great deal of difficulty comprehending the magnitude of what is going on at this climactic moment of life's history on earth; most of us are caught in the solipsism of human supremacy that is reinforced everywhere by a man-made, or changed, environment. Ecological restoration and wilderness preservation are, finally, the politics of *Homo sapiens* belonging to earth's ecosystems. We're all in this web together, and our very human nature depends on that quality of richness and complexity in the living tissue that enfolds us and sustains us. "Evolutionary diplomacy," Freeman House calls it.

Accordingly, habitat restoration and the reintroduction of extirpated species become important human occupations, tributary to this larger process of giving evolution room to move. Development of techniques by which to do this—to safeguard, for now, the genetic endowment of native species wherever they are, to rescue and propagate them wherever they are—is a transformational practice. It sits within the larger cultural and political task to which Aldo Leopold alluded in remarks made to the Wildlife Society in 1940: "To change ideas about what land is for," he said, "is to change ideas about what anything is for." It means changing our ideas about what *we* are for: is ours to be a dominant, parasitic relationship with all other species, or is it to be coordinate? It's the big ontological question, and we can't balk it.

For restoration activity to be honorable, it must be inseparable from the prevention of further habitat destruction. Otherwise it misrepresents the founding facts of specificity: that places are different and that plant associations, which are the basis of life and livelihood for most other organisms, ourselves included, arise where they belong and belong where they are. Microclimate, microhabitat, differences in soil composition—all give rise to genetic variation even within species. Thus an ecosystem is a firmly rooted thing. The possibility of certain kinds of ecological restoration—establishing new patches of prairie plants, for instance—should not lull us into the delusion that ecosystems can be moved around like oriental rugs. Nor should we make little plans. The top carnivores and omnivores that are the truest sign of an ecosystem's health and wholeness are cosmopolitan and wide ranging. Thus their main habitat requirements may be freedom from human harassment and room to move. What must be restored to ensure a future for the wolf, the grizzly, the mountain lion, and other great predators is *extent*.

If we fail to preserve and restore, we may default into an ersatz world where people have to get their kicks in virtual reality because earthly reality has become desolate. In assenting, however unconsciously, to the destruction of biological diversity, we domesticate ourselves and constrict our experience. Wild and free are much the same and the antithesis of domestic. Wilderness preservation and ecological restoration are, at different scales, homage to undomesticated creatures and communities. They hinge on authenticity and indigeneity. Prairie restorationists in Chicago try to limit their gathering and propagating of plant matter to ecotypes—populations that can be found within a fifteen- or twenty-mile radius of the site being restored. Advocates of wolf reintroduction in the Sonoran bioregion consult federal records on wolf sitings so that their proposals for migration corridors will be authenticated by natural history. Salmon restorationists in northern California work for a race of fish native to their river, not for hybrids, clones, or next best.

This loyalty to the minute particulars of DNA in native species stands in contrast to the genome craze of our time. Conservationists and restorationists are interested in genes, as they are spontaneously and integrally expressed in organisms, for the fact of their being. Genetic engineering continues the mania for control and exploitation, the conquest of the tiniest redoubt of the wild. For genes are wilderness, too: submicrocosmic wilderness, made of, by, and for whole places.

One winter night I had a conversation with a friend in which we were both blundering around, lacking sufficient philosophical method to demarcate the boundary between natural and man-made. As making this distinction challenges even the bigtime pro philosophers, it is not surprising that we amateurs wandered so bootless.

My friend was trying out the proposition that everything that is, is, in a sense, a product of evolution—that virtual reality *is* reality and that bionic limbs and computer-chip replacements for areas of the brain, by dint of functioning as part of an organism, will come to be regarded as living. I ransacked my logic for arguments against this sacrilegious proposition; such artifacts are not alive because they lack the capacity for self-healing, because their manufacture entails pollution, because they have no emergent properties, because, in the case of computer chips in the brain, they will bring forth no surprises. Nothing in a category that has not already been thought of will be conceived by them.

As the conversation forged on, I was privately, uneasily, aware that my revulsion toward the whole realm of robotics and cybernetics, toward virtual reality and artificial life, has left me willfully ignorant of some doings that may well blur forever the distinction between natural and man-made. There are roboteers who think it would be an advance to replace our brain tissue with computer circuits. There are researchers working on "manufacturing" processes that could take place at the atomic level—nanotechnologies—and may amount to a kind of engineered metabolism. And there are engineers and scientists devising machines that can propagate and repair themselves, computer software that "designs" organic, unpredictable forms.

The manipulation of the genetics of organisms in order to maximize one or two of their qualities, like milk production in cows, or frost resistance in strawberries, or antibody production in cells, is proceeding at a brisk clip toward breaching the sanctity of life. That the term *life* is now applied to things apart from biology is a measure of how rapidly this audacity has proceeded.

Such hubris spawns bad accidents that spill out and engulf innocents—creatures quietly minding their own business, living out their evolutionary destinies, each having an integral relation to the whole. I worry about any optimism that tempts us to become careless about all these other actors of fate. Casually we propose to go forward, to go nature one better, to design plants immune to pests, animals that produce like factories, machines that mimic cells, and people invulnerable to death. The human mind can be lethally self-interested: "autistic" is how culture historian Thomas Berry has characterized the postwar generation.

It shakes out as a religious question, I guess. And the question is: Is nothing sacred? Are there no natural phenomena—cells, organisms, ecosystems—before which we might just stand in humble awe?

As our conversational play with abominations and holies wore on, I became short-tempered with my friend. Finally I withdrew, frustrated. I had not been able to forge a purely intellectual argument for the sacredness of nature and the profanity of industrial civilization. Nature per se—ancient forests, prairies, termite mounds, and hollow wing bones—is the argument for humility. Why this is not self-evident eludes me, but I guess it has something to do with the shape of my yearning for the holy.

Much pleading by ecological activists in the twentieth century has urged us to let trees fall in the forest (it matters not whether soundlessly) and rot—for the sake of the soil, for the sake of *that* community. It is a plea that the value of nature is intrinsic and need not, indeed should not, be accounted for only in terms of what humans derive from it.

This is a hope, not a theory. If increasing numbers of us participate in ecological restoration where we live, we will learn the difference between natural and man-made. Through positive engagement in ecosystems, we will come to revere evolution's fine and canny design processes so greatly that we will stop at nothing to preserve all remaining species and the habitats they need.

What I want restoration to be for *in us* is the transformation of our souls. In addition to what it may accomplish in the land, I yearn for it to be the yoga that will cause us to evolve spiritually. That is to say: humble us, and restore to us a capacity for awe in something other than our own conceits. I fervently hope that this work—in our streets, yards, fields, woodlots, parks, creeks, and watersheds—does indeed hold the potential to carry us into a postmodern, postindustrial, postcivilization pattern of relationship with nature. I pray that it is our way back into the web of life, carrying with us what we know now.

We cannot wish ourselves ignorant of technology and science, of specialization and comfort, of culture and abstraction. But in our efforts at repair, as we are baffled by the ungovernable processes of time, space, and Gaia, we may learn to assign the creatures of our mentality a lesser value and to subordinate our infatuation with technology to our reverence for the mystery of life.

We can behold that mystery in a frothy orchid rising from the forest floor, in the grotesque wit of a walking stick insect's camouflage, and in the fact of our dying. It is in the implacable elaboration of the beginnings, endings, and perpetuations of millions of species throughout the biosphere. To confront this wild unreasonableness of nature is to look god in the eye and relent. It is not to relinquish consciousness, but to locate it in an organic whole. It would mean to belong, to participate in nature, rather than taking it apart. It is in our souls that this epoch will be resolved, but our souls are engendered by the universe through the biosphere. We cannot know the good, the true, the beautiful—we cannot know what is right—in absence of nature. The solipsism of value systems premised solely on human ideas is writ large in havoc around us. It is time now to change. We must, finally, come to love and respect that which cannot be owned—and to cherish it fiercely.

———

Talking with my Wisconsin friend Roger one July morning in 1993, I told him that, on his weekend tour of Leelanau County, he should bear in mind that every place he'd be going—fields and farms and open country—there would have been old-growth hardwood forest with a few six-foot-diameter elms scattered through it. In this conversation with Roger, who is, among other things, a farmer, I gained an appreciation of the pioneer's investment of toil in clearing this country. I got a sense of the muscle in the heroic saga of opening ground for agriculture. It is not so difficult to understand why growing food should be felt to be unequivocally good, and why the price paid for that in human sweat should be honored. Agriculture and its resultant civilization have, for millennia now, been upheld as offering a greater good than savagery in the wasteland. So these patterns not only dominate our economics and culture, they constitute much of the structure of myth.

I've always chafed at the idea that a new myth is required for us to make it into the next millenium without razing the biosphere. The deliberate creation of an authentic myth seems so paradoxical. But this morning's insight suggested myth is certainly one of the dimensions where restoration must work. The Gnostic author Stefan A. Hoeller proposes that "the restoration of biological order must be seen as a metaphor for the restoration of the wholeness of consciousness; otherwise we shall miss the deeper meaning of the present crisis of civilization."

Healing is the dominant metaphor in ecological restoration, as in so many realms of the current human drama. Restoration's leading exponents and theorists speak in those terms. Some of them even hint that the healing process is as much shamanic as scientific—that a ritual sense should inform the work. This might become a way of dealing with our all-too-human problem of impotence and dominance in the face of nature. Dominance has been encoded in our thinking about everything—from how we treat one another to how we treat the living world. Cooperation, attentiveness, and partnership (sometimes even with unseen forces)—the values that prevail in restoration and shamanic healing and effective community—are perhaps more difficult to mythologize because they cannot be drawn in terms of individual heroics or victorious battles. Our tutelary images could, once again, come from nature: plant and animal powers guiding, allying with, and correcting human conduct.

Most people on earth are emigrants and immigrants, more or less recently. Lately the pace of deracination and the urgent need for new home places have intensified. Perhaps this quantitative difference equals a qualitative difference. It spells an end to indigeneity and a need for new eras of becoming native-again-to-place, of reinhabitation. Restoration needs to be considered one of the ways and means of reinhabitation, along with ecological agriculture, renewable energy development, and self-governance—all crucial to the future of nature.

As I have worked on this subject, something that I originally envisioned as a minor element—environmental history, nature's past—became at least as vital as restoration. Retrospect engrossed me for the better part of a year. The past of the land and of its people discloses its true nature and suggests the character of an autochthonous future. That late-twentieth-century society might course-correct and rebuild its communities in tune with the rhythms and cultures of definite places seems less far-fetched to me than the idea that another bureaucratic technofix might be made to work.

During this era of civilization our cultures have become increasingly effete and dependent on artifacts—political and technological—which afford rude survival for the many and luxury for the privileged few. I think we will be weaning ourselves from this infrastructure over the next few centuries as we explore what it might be to be "future primitive." The fact that the human population is now so immense, and that so much of the planet's habitat has been impoverished biologically, does not mean that we can't or shouldn't imagine a future for ourselves in our ecosystems, maybe even a nomadic future, a wild post-civilized future. It just means that it's a few generations farther down the road. We won't live to see it fully realized, but even just beginning, it feels good.

My ultimate hoped-for-outcome from restoration is not an unremitting opportunity for high-minded horticultural toil but an eventual return to an ecosystem balance (which implies a much smaller human population) and hunter-gatherer proportions of leisure and dream time—time for the kind of social play that results in a diversity of cultures as varied as the ecosystems that sustain them.

Accordingly, one rubric for right action might be to ask whether my work today is contributing to the restoration of natural and cultural diversity toward an elegant subsistence for future primitives. Is what I am doing contributing to the knowledge individuals require to live independently and harmoniously and harmlessly on the Earth, or is it leading to more sameness, more dependency, less joy, fewer species?

Community can grow, embedded in restoration endeavor, galvanized by wilderness preservation struggles, fostered by concerted human action in service to the health of the land, one's range on the living planet. Barn-raisings made people feel good a century ago. Salmon release and prairie burning do so today. Who knows what restorations of joy there might be a century hence? Buffalo runs down Vincennes Road in Chicago? Soil, and tigers, and native rice again in Tamil Nadu? Grizzlies from Mexico to Newfoundland? Waterfowl rising from the upper Mississippi tributaries and potholes to darken the skies and weigh down a few tables? If we are writing the future of nature, as the oracle Amalamu counseled, it had better be bright, no less prodigal than earth's own genius.

The Politics of Wilderness
and the Practice
of the Wild

R. EDWARD GRUMBINE

Smoke in the chill gray air.

We climbed up the steep road from the coast accompanied by a change in the weather and an onshore breeze. In the center of a jumble of cinderblock buildings, vintage house trailers, and rusting hulks of dead automobiles stood the wood slab roundhouse. Weedy blackberries rambled nearby. At first, not knowing back from front, we couldn't find the door. But soon the entranceway, a passage constructed on the north side and filled with firewood, drew us in.

Upon entering the roundhouse, the Kashaiya Pomo of north coastal California turn in a circle clockwise. But if you're a first-time visitor, they don't expect you to follow this ritual. Nor do they come right out and tell you about it. If you don't detect it for yourself, it might never be explained. So it's good to pay close attention and not hesitate to inquire about appropriate roundhouse behavior.

As our eyes adjusted to the light from the fire pit near the door, we could see about fifteen people seated on rough benches that curved around the inside walls. Crackle of fire and murmur of voices, people were saying hello, visiting, gossiping quietly.

We looked for our friend, Lorin Smith, who had invited us to the Fall Acorn Festival. Lorin is a retired state park ranger and the spiritual leader (*yomta* or doctor) of the Kashaiya people. He was sitting on a special chair (special because it was the only seat in the house with back support) against the east wall. He greeted us and we visited, after placing our potluck food on the tables in the rear. Other people stood and ate, as the kids eyed the desserts.

The Acorn Festival is the most important public ceremony that the Kashaiya Pomo conduct. Anyone may come to celebrate the harvest and the beginning of the winter rainy season. Over the next few hours, people continued to drift in. Most of those present were Anglo. There were perhaps ten to fifteen Pomo.

I noticed the centerpost. It was massive, over four feet thick from the hard-packed earthen floor up to the roof beams. It was smooth and dark-cured from years of smoke. Lorin told me that the elders found it in an old clearcut left for slash. It was hauled to the new roundhouse site by truck, but the Pomo raised it in place by hand.

The ceremony began without any obvious sign. A Pomo woman stood up and greeted everyone. There were thirty or forty people in the roundhouse now, though one hundred and fifty could have fit in with room to spare. The woman invited everyone to join in the singing and dancing. Though there was little sense of formality to her remarks, it was clear that the Pomo were following certain procedures.

A chorus of male and female singers sang repetitive lyrics over the beat of clacker sticks. (The Pomo do not use drums.) Smoke curled around the centerpost and flew out the smokehole. Clacker stick sounds rattled up to the roof beams. Some of the songs were accompanied by dances. At times, men and women would dance together; other songs were danced by the women alone. I was surprised that Lorin did little singing and danced only once. But for this he put on a fancy western shirt, held oak boughs in his hands, and danced solo.

The Pomo accepted us as neighbors and friends. Songs were shared and dance steps were explained briefly. During the dances people talked, sang along, moved around a bit, and went outside to pee. It wasn't like being in church or sitting Zazen. After some time I was less sure of where I was, what kind of person I was supposed to be, and when events were supposed to begin and end. But though my boundaries were being stretched, I was not concerned. The centerpost held up the world and it was firm, stout, and round.

The ceremony ended with Pomo and Christian prayers of thanksgiving: to acorns, the forest, the fire, the roundhouse, to the singers and dancers, to Great Spirit, and to God. We said good-bye and drifted out the door—cold night, waxing moon, and thousands of stars dancing across the roundhouse of the sky.

Later, I couldn't get over the fact that the chorus was composed mainly of white people. Most of the participants were Anglo, too. The roundhouse sat in the middle of a postage-stamp reservation surrounded by non-Indian cutover timber company land. The Pomo had been so accommodating, so gracious. But they were hanging on by the skin of their teeth—the rusting trailers and junked cars proved that. What did the people do for work? With so few Pomo participating, how could the songs and dances survive? Did the people have some strategy for survival or were they merely resigned to the overwhelming presence of white voices singing their songs to keep them alive?

———

A month later and fifteen hundred miles away, I sat on a comfortable sofa by a roaring fireplace in the meeting room of a guest ranch in the heart of the Greater Yellowstone Ecosystem. Snow lay on the ground outside and daylight faded against the Gallatin Range. Next to me on an end table sat an expensive cast-bronze of an Indian warrior on horseback. I was a member of a group of environmental activists gathered from around the country to discuss forest reform strategy. What might be done to increase pressure on Congress to protect wilderness areas and biodiversity?

I had been asked to explain how ecology and conservation biology might provide new strategies to aid the cause. First, I described why setting up a nature reserve system in the United States had failed. Parks were too few and too small to prevent extinctions over time. Political dealmaking could not encompass the biological needs of species and the ecosystems they depended upon. Second, I outlined how conservation biology principles could be used to expand current proposals for new parks and wilderness. Science could also be summoned to argue for Endangered Species Act reauthorization. I proposed that the goal of saving wild places might be replaced with the goal of protecting ecological integrity. This new goal would include maintaining and restoring viable populations of all native species, representative samples of all U.S. ecosystems, and ecosystem patterns and processes. I concluded by suggesting that if ecological integrity became the new watchword of the environmental movement, we might be able to protect many more acres of wildlands than if we focused only on roadless areas.

During the talk my thoughts strayed outside. Hiking up the Yellowstone River the previous day I had seen a bald eagle fishing the shallows and a dead elk whose hindquarters bore the marks of Coyote scavenging for a meal. What if, somehow, the people in this group could sway the political process so that the goal of ecological integrity might become manifest? What if a hundred greater ecosystems as healthy as Yellowstone returned to life across the continent? Why not a Greater Tall Grass Prairie in Minnesota? A Greater Northern Forest Ecosystem stretching from the coast to the interior? A Greater Gila-Cochise Ecosystem in the Southwest? A Greater . . .

I felt energized as my talk ended. But the group's response astonished me. Though many appeared to be in general agreement, several participants raised challenging questions.

"With all this emphasis on biology, where are the environmental ethics in your position? I don't see protecting wilderness as having much to do at all with what you've just said."

"What is your standard? What do you believe is worth fighting for?"

"This science is all fine and good, but do you believe in wilderness or not?"

Flustered, I attempted to explain that the goal of protecting ecological integrity was shot through with values. And that, using the Greater Yellowstone Ecosystem as an example, more lands might be protected if activists looked beyond the remaining pool of roadless lands and instead based their strategy on the biological needs of grizzly bears. But my response was hardly compelling to the group. We broke for lunch.

Months later, I am reminded of the roadhouse, set among cutover forests and decrepit government housing, and how what happened with the Pomo is related to the politics of wilderness and the prospects for protecting biodiversity. Pomo behavior during the Acorn Festival offers signs of wildness and hope for the future. There were several tokens present that night: fire, darkness, an earthen circle in the round. But I had experienced these before and in wilder places—campfires under the great western sky out in the back of the beyond. What impressed me were signs that we label "cultural" but in fact are also "wild." The Pomo constructed time so loosely that I felt at times like I was backfloating in an immense ocean. The evening did not so much progress as unfold. The most important ceremony of the year was conducted as informally as I might brush my teeth or sweep the kitchen floor. There was, however, focus to events: the fire was fed, songs began and ended, people danced certain steps, the earth's bounty was remembered. The round-house created a boundary and also brought in the world outside so that a great intermingling was possible.

But the Pomo are not flourishing. There was no acorn mush to be eaten at the Acorn Festival. The Pomo have, quite literally, lost their major source of nourishment. Words to songs were forgotten too often. This was an embarrassment to the Pomo, not a product of informality. Most troubling was the lack of Pomos participating. The roundhouse's celebrants were mostly non-Indian. This was acknowledged by the lead woman singer at the ceremony's close: "You people here are helping us, you know, to keep the songs and prayers alive."

And the Greater Yellowstone participants were helping to keep diversity alive. But unlike the Kashaiya people, the group gathered at the ranch had been socialized to separate culture from nature and people from places. Evidence of this culture/nature divide is demonstrated by the questions at the meeting. The Yellowstone activists could not break with the wilderness tradition that depends upon pristine nature as a source of human inspiration—in contrast to developed lands devoid of any wild qualities. That tradition, as worthy as it has been, rests upon one of the key assumptions of history: culture can only be opposed to wildness. This assumption is not inconsequential. It allows most citizens of industrial societies to inhabit a world that is, by any reckoning of the facts, being destroyed by industrial imperialism. Wilderness reserves and protected areas are expected to offer a balance to environmental destruction and escape for the weary urbanite. Yet many activists, who do care enough to fight industrial power, unwittingly maintain the people/nature fence. They wish to perpetuate, instead of extend, the wilderness philosophy and tactics inherited from nineteenth-century Romantic ideals of the balance of nature. But this balance has never existed.

Consider the terms *biodiversity* and *sustainability*. Where have they come from and why have they appeared today? The answer is that they have appeared today because they are disappearing so rapidly. Conservation biology has grown in direct proportion to the increasing rate and scale of extinctions and habitat destruction. Debate over what is sustainable behavior has emerged as people confront hard evidence that such behavior no longer exists in many places. The biodiversity crisis is forcing Americans and others to reevaluate not only their political commitments but also the cultural assumptions upon which these strategies are based.

During my time with the Pomo, sustainability and wildness were never discussed. In Yellowstone we talked incessantly about these two matters. But we could not agree that the American idea of wilderness is an article of faith born of the very power that is destroying wildlands. This same power favors balance over uncertainty and surprise, *Homo sapiens* over all the diversity of life, brotherhood over sisterhood, one dominant culture over multicultural variety, economic well-having over ecological well-being, and the politics of wilderness over the practice of the wild. The radical Western split between nature and culture allows us to presume that ecological sources may be transformed into natural resources for human use only, that a human house cannot be round and wild, and that a mountain lion's home range is immaterial to the dedication of wildlands. But what if natural resources don't exist? What if legally defined wilderness areas are figments of our cultural imagination?

———

What is *wild*? Wildness escapes easy definition. While wilderness in America is most often a place, wildness is the force behind places or, as Tom Lyons says, "the overarching reality that transcends all plans and creations." A biologist might study natural selection and adaptive behavior in the field, but the process of evolution is wild.

Humans, too, are wild. Wildness in people might be characterized as the self-regulating aspects of *body* interacting with the unconscious depths of *mind* with each of these in constant contact with *environment*. In the roundhouse, potluck food graced the tables in back. Acorn boughs were laid against the walls. Humans sang and danced. Neighbors were recognized, received, and respected. Wildness was flowing without legal proclamation.

Classified wilderness areas may allow wild nature to live and breathe to the extent that these places are less subject to human control. Being in wilderness gives us the opportunity to connect with wildness and collapse the culturally relative categories of modern existence. Wilderness and wildness intersect where a river, mountain, elk, or Indian paintbrush spark awareness in us that helps to break down the fence between people and nature.

But the fence between culture and nature is difficult to reduce. It springs out of a fundamental paradox of human existence: the distinction between self and other. This boundary appears to go beyond history at least as far back as the beginning of the Neolithic era. The Pomo recognize it; so do all modern peoples. We cannot deny the distinctions between humans and other species, coyotes and spiders, or any member of earth's community of life. But we can decide what to make of these differences. Up to the present day, our choices have been poorly made. When confronted with otherness, we have chosen opposition over accord. Taking conflict as our standard, we have separated culture from nature and wilderness from wildness. To paraphrase writer Barbara Allen, we haven't lost our relationship with wild nature, we have simply invented it in terms that do not allow us to sustain cooperative relations with it.

In adapting culture to fit the "rules" of wildness, humans may choose between fear of nature or solidarity with it. But these choices are never cut-and-dry. They are always tangled up with such ongoing human predicaments as the need for individuals and groups within societies to come to terms with answers to the questions that cultures have already imposed. Even if every wilderness activist replaced the idea of wilderness with the goal of protecting ecological integrity, Congress and the general public might be unwilling to go along. Not every Pomo who lives at Kashaiya associates with the roundhouse. Each individual must furthermore define through personal (and cultural) experience just how wide the boundaries of the self may be extended toward others. Here the deep ecology movement has offered wilderness advocates many powerful examples. But answers to these questions are always provisional. In an evolving world, ecosystem edges are always in flux and human borders are continually subject to negotiation.

———

How might we begin to integrate wild culture with wild nature? Ironically, the first step in protecting wildness is to continue to focus on increasing the size and number of protected areas—that is, wilderness. These lands are the last surviving remnants of wild diversity and they are faced with imminent development. But protection of biodiversity should be emphasized over preservation of scenic lands and recreational opportunities. Given the politics of wilderness protection in the United States, the best hope for success will be to ground such an approach in science. A conservation biology platform would include: habitat protection for viable populations of all native species; areas sized large enough to encompass natural disturbance regimes; a management timeline that allows for the continuing evolution of species and ecosystems; and human use integrated into the system of protected areas that would provide for *Homo sapiens* within the foregoing constraints. The hope of protecting large wildlands is that this strategy would slow the rate of the biodiversity crisis while also providing people with experiences that could feed the nature/culture system both ways—sustaining wildness at the core of protected lands as well as at the center of human communities. The promise of this strategy is that as humans begin to gain direct experience with ecosystems by working to protect biodiversity, wildness may explicitly become part of culture again.

———

Deep in wild mountains, a ghost bear burrows into a hillside for winter's sleep. She has never encountered a human being, so people say her kind do not live there.

In the slow-moving shallows of a Georgia river, a few freshwater mussels congregate on the backside of a boulder. They are endangered, the last of their species, and their stretch of stream is about to host another new tract of suburban houses.

In the roundhouse, each fall people still sing and bless the oak trees. They dance and invite neighbors to the feast. Sometimes the singers forget the words and the dancers must stop to remember steps before they begin anew. I don't know the words to the songs or the steps to the dance, but I pay attention as best I can, hum along, and shuffle to the beat. For the Pomo, as for many people today, the fit between culture and nature appears unsustainable. But the centerpost, deep in earth and shaped by human hands, holds up the roof of the world. The great sky is just outside. It is inside, too. I can see it through the smokehole.

Looking Salmon
in the Face

NANCY LORD

Summers, I live in a wild place, what we in Alaska refer to as "the Bush." This particular coastal area isn't designated as wilderness, set aside, or protected. It's just that, for now, few people have reason to be here. My reason, ostensibly, is to fish. June through August, Mondays and Fridays, my partner and I catch and sell salmon that pass our beach on their way to spawning streams. The rest of the week, and parts of May and September, Ken and I mend nets, comb the rocky shoreline for log poles and splintered squares of plywood, do a thousand camp chores and projects. We live quite happily in a tiny cabin at the top of the beach, among the eagles and mosquitoes, our growing collection of agates and circle rocks strewn everywhere around us. We put up our own supplies of smoked and canned salmon, and, from the middle of July on, our teeth turn black with blueberry stain.

The fishing here is poor, for complex reasons having to do with fisheries management and increasingly efficient fishing fleets, and other fisherfolk have abandoned these shores for more lucrative spots. In ten miles of coastline, we have only three remaining neighbors, all old-timers a generation or more beyond us, men too old to begin fresh anywhere else. Stubbornly, we stay because making a living is less important to us than living where we want to be, in a spare, uncrowded place.

In an unlikely reversal of most of America's expansive history, there are probably fewer people living along our stretch of coast today than in any other historic period, fewer perhaps than at any time since the first

Native people arrived through the western mountains some six hundred years ago.

The life of this place has always been closely tied to fish, particularly the five species of salmon that migrate north through the course of the summer: king salmon beginning in May, then reds, pinks, chums, silver salmon trailing on into the fall. The first people here, Dena'ina Athabaskans, caught salmon in weirs and with snares and dipnets made from spruce roots. They smoked the fish with a cold fire of green alder or cottonwood and cached bales of it for winter food. Dena'ina writer and scholar Peter Kalifornsky, who died in 1993, told how his people traditionally followed the fish runs, catching and preserving what they needed for winter supplies. "It was the rule that one day's allowance of food was a piece of dry fish as big as from the meaty part of your palm at the base of your thumb to the tip of your middle finger." Salmon were central to Dena'ina culture, its very life source.

When the salmon runs failed, or the Dena'ina couldn't catch what they needed, or the winters lasted too long, the people starved. Kalifornsky told of a famine long ago at this place. "What they had put up for winter was gone. There were no moose or caribou. Black bear and porcupine were all that was left. It snowed repeatedly." According to the story, two men went to look for something to eat, along the beach flats toward the river. When they stopped to drink at a slough, they fought over a last bit of dried fish, and one drowned the other. After that the slough was always known in Dena'ina as "Where Someone Shoved a Person Under Water."

Later, the Russians introduced fishing nets, and for a time it was easier for the Dena'ina to catch what fish they needed. It wasn't long, though, before Americans and their canned salmon industry took over. By the late 1800s fish traps scooped up most of the fish for canning and export, and the Dena'ina had to compete for both resources and wages with hundreds of transient fishermen, cannery workers, and others drawn to the region. Some sold fish to the canneries; others worked in them or at jobs such as cutting poles for the construction of fish traps. The numbers of Dena'ina, which had never been great, fell rapidly. Smallpox in the 1830s had already cut the population in half, to 750, and a flu epidemic hit hard in the 1920s, decimating entire villages.

A village near here, which still shows up on many maps, is one of those decimated places. Once a substantial settlement on a bluff overlooking mudflats and long sculpted sandbars, it was abandoned in the late 1920s. In spring, I've walked over the village site and looked through the mat of flattened dead grasses at the rectangular outlines of the old houses—the earthen rims, the hollows, all that remains after desertion and fire. I've thought about what it must have been like to live there in winter, in deep snow, with thick jumbles of ice piled up on the mudflats below, to have heard all the stories and known all the names. At one time there was a Russian Orthodox church on the hill, and an American cannery at tidewater. Today, there's a single summer fish camp on the beach, one of our few neighbors who, like us, boards up his doors and windows at the end of summer and heads for somewhere else.

On our beach, only on the lowest tides, we can walk out among a line of stumpy wooden posts. Sheared and rounded by decades of shifting sea ice, these pilings are all that's left of a huge fish trap that once helped support the nearby cannery. Elsewhere at low water we find the remains of wooden and corroded metal stakes used by fishermen before us to secure the outer ends of their nets. They remind us always of the human history of this place, all the years that fishermen and fisherwomen, Native and white, worked this beach. Our neighbor, George, who's fished here since the 1940s, talks of them still—Nickanorka, Hanny, Koch, Haynes, Hightower, Doris, and Chet. More than once there were fish wars—fighting, with fists and guns, over fishing sites. George one summer hired huge crews, entire families, and set them up all along his miles of beach with skiffs and nets to hold off the men he called "poachers," site jumpers.

All those years, frontier years before, during, and after statehood, people came and went from here, fishing, trapping, even trying to homestead in the country above the beach. One man dragged building logs through the woods with a horse, shot the horse when he was done with it, but never "proved up" on the requirements that would have given him ownership of the land. Chet and Doris, twenty-plus years ago, were the last year-round residents, fishing on the beach summers, moving up over a trail at the end of the season to winter in a cabin on a lake. Doris was Dena'ina; Chet was a white man who, George says, "got more Indian until he forgot even how to write his name." There was a boy who ran away on the salmon tender after his father shot at him, and there was Slim, who used to eat out of the same pot as his dog, even among company. Young men drowned in the cold, cold water. Just last September two skiff travelers stopped at our neighbor's. The next day, they tried to put ashore in a bay farther south, rowing a dinghy from their moored skiff, and they overturned. This spring, when I opened George's guestbook, I saw their names, the last two loopy signatures in the book; George had penned in beside them, *drowned*.

There was, of course, all these years, a relaxed, rivalrous, or confused cohabitation among the people and wildlife of this place. It was a time of *good* animals and *bad* animals. The good, like salmon and beaver, were taken for their commercial value. The bad were simply killed; they were dangerous and destructive, like bears, or competed for salmon. For thirty-five years, until 1952, the government paid a bounty on eagles to reduce their take of salmon; no one seemed to notice that eagles fed primarily on spawned-out fish. Bounties were paid on seals, too, by the nose, and on Dolly Varden, a sea-run trout that eats salmon eggs and fry. When biologists finally analyzed the Dolly Varden tails that were being turned in for bounty payments, they found that most belonged not to any trout at all but to silver salmon.

Today, up and down the beach, there are long distances between fishing nets, an easy neighborliness. Empty cabins are boarded up, sliding into the sea. Salmon are still the reason for human occupation, although the relationship between fisher and fish has changed, is always changing. The fish go to market, first pitched from our skiffs to a tender, then carried on ice to processing plants across the inlet. They're not canned anymore, but cleaned and headed in a style called a "princess cut," and frozen, and shipped to Japan. Ken and I, like other fishermen, think of these salmon less as food than as product. We handle them like sticks of wood, snapping them from web, pinching gill plates for grip, dropping them to lie—gurried and warm—in the boat bottom. Our attention goes to the differences between species—scale size, spots, the calico chum—that tell us how much money we're making. The high-value red salmon we add up as five-dollar bills.

Our relationship to those salmon we eat—the ones we set fresh on our cleaning table and drain of blood—is something else. I slit their throats, and bright blood pulses out, drips like thickening jelly into the sand. When I rinse them with the hose, the bloodless fish react by twitching and tossing, nerved attempts to again reach saving water. I look at them, elegant silver salmon with scales like sequins, backs blue or gray, sides rounded or flat or limned with a hand shape, the almost-grasp of a seal's paw. Sea lice stick in resistless hollows just fore of the tails, and the gillnet has scored the back of heads. The pectoral fins are stiff and flared, and still quivering. A silver-streaked tail is nicked. I look into salmon faces, eyes, gaping toothed mouths. This one is snouty; that one has a clear, wet, perfectly ringed pupil. This one has fleshy cheeks, the other small, china-doll features. Every fish face— every fish—is different, individual, worthy of recognition.

"Everything has a life of its own," Peter Kalifornsky taught. Everything. Not just people and salmon and gnats, but every tree, flower, rock, grain of sand, every wave, cloud, raindrop, shift in the wind. The traditional Dena'ina, of course, lived more closely with the rest of the participants in their world than anyone today. They depended for all their food, medicines, all their material culture, on what was at hand, and they developed elaborate rituals and belief systems to guide their relationships. Every year, when we remove the first king salmon bellies from our smokehouse, we gorge ourselves on the fattiest parts and feel some small part of tradition: the seasonal cycle of plenty, renewal. But the next day we make lunch from grocery store food delivered by the tender, avocados shipped thousands of miles. We know we'll never starve, just as we'll never treat sickness with pulverized wormwood leaves or replace our aluminum skiff with one made from seal skins.

In our 1990s lives in this place, we don't need porcupines to eat, seal gut to shield us from the rain, or beaver furs to warm us. We don't need beluga fat for preserving clams, or sinew for sewing. Fiddlehead ferns unfurl and the fireweed toughens on its stem; we rush to pluck neither in its prime, and neglect to hunt for salmonberries when we have fruit cocktail in a can. Our boat will motor against the wind and tide, and we set our nets by the clock, by law and regulation. It's easy, in this way we separate our lives from the others around us, to forget sometimes to acknowledge them, even to recognize their presence. When we forget to notice the rockweed thickening on the reef rocks or the inky cap mushrooms poking up overnight beside the trail, it's because we aren't looking them in the face.

These days, harbor seals boldly, unstintingly, keep watch over our nets. They race us to each gilled salmon, even as we shout "Catch your own fish!" and bang an aluminum post against the boat hull, imitating a rifle sound they have likely never heard or learned to fear.

Beluga, the only whales with flexible necks, turn their heads toward us as they pass at low water, just beyond the end of the reefs. All those porpoising white backs make the flat water look like it's risen into whitecaps. They're supposed to sing like canaries, but all I've ever heard is the breezy whooshing of their breaths. One day they're spread out, dozens heading north together, and the leaders slow and circle off our point, waiting for the stragglers. There are perhaps a thousand in the inlet, and each one is said to eat sixty pounds of salmon every day.

Mornings we read fresh bear tracks in the sand: a solitary black, a large brown sow with a cub, a pair of small browns that stopped to skirmish. Less often, one walks past us in daylight, so like us, so familiar, it's easy to see how peoples all around the world have considered bears kin, half human, keepers of a parallel culture. A golden bear snarfs the tideline of clean spawned-out hooligan. Black bears eat watermelon berries in the backyard and bite the roll of toilet paper in the outhouse.

Just as the smarter, more aggressive fishermen are elsewhere, the larger, more dominant male bears are along the rivers, rolling in fish. The bears that share our summer space are on the territorial margins— blacks squeezed by the bigger browns, adolescents, the less fit. Teenage black bears, like their human counterparts, have yet to learn judgment, anything about their own vincibility. They come to the cabin door and only *hrumph* and stare shamelessly when we shout and pound cooking pots together. It's not so much that we fear them as we fear *for* them. Our less tolerant neighbors shoot to kill.

Beavers go about their business in the country above the beach, damming creeks, building more and larger ponds we have to divert our trail around. They drop birches two feet thick, open dark forest to sky, redirect the flow of water down the bluff. When we row on the lake, they swim out and circle us, curious.

A bald eagle sits on a boulder out front and waits expectantly. George, our neighbor, calls this one Grandpa, lays out fish heads for it every day, has it almost feeding from his hand. Ken and I always dump our fish carcasses from the boat, or off the end of the reef, to be swept away. We don't want to attract bears, or teach dependence, or encourage seagulls. Seagulls are the disgraced animals of the nineties; they thrive on fish waste, on garbage dumps, and in increased numbers they overwhelm seabirds, eating up their eggs and chicks. We don't put out food, but we do put up a birdhouse. Blue-green swallows nest, raise a brood, fly off in midsummer.

Driftwood stacks up at the top of the beach, huge cottonwoods with layered bark, spruce etched with the tracks of beetle larvae. Agates accumulate among the sand and gravel. Long silences give themselves to the far-off calls of loons. In August, a full moon rises over the inlet and is huge and fiery, reflecting, lighting the night. Wind blows the grasses flat, and fireweed lets go its fluff like snow. Rocks shift in the creek, and the muddy inlet continues its incorrigible widening and filling, its geological journey.

Everything has a life of its own, but nothing lives by itself. And nothing stays the same.

What are the new rules? How do we live with one another—seals, salmon, porcupines, maggots, chickadees, sea lice, monkeyflowers, moonrises, plastic buoys, people? I don't mean law, the rules of the Marine Mammal Protection Act, the Endangered Species Act, refuges, sanctuaries, licensing, Corps of Engineers permits. I mean: how do we live with one another, by what moral authority, what tribal understanding? Some well-intentioned souls would drape laurels around the bears and the whales, would forget the hooligan, the mosquitoes, and especially the people. The challenge, as I see it, is not to try to understand one thing and another, isolated pieces of information, but to embrace the culture—the entire ethos—of wild and all other places, wherever it is we live. We can do this the way people always have—with rites, rituals, traditions, lessons, taboos, the sets of stories we tell ourselves and each other, time after time.

To begin, it helps to look closely into every face.

Reference

A Dena'ina Legacy: The Collected Writings of Peter Kalifornsky. Fairbanks: Alaska Native Language Center, University of Alaska Fairbanks, 1991.

The Reflective Ebb and Flow of the Wild

ALAN DRENGSON

Recently a friend and I ascended the summit of an unnamed peak in the heart of the Olympic National Wilderness Park. I had been to this summit over forty-five years earlier. On our way up I noticed how little the immediate area had changed over those years. That gave me a sense of security and deep connection. We reached the summit and sat looking around in all directions, horizon to horizon, up at the sky, down into the valleys. What we saw cast a chill on my heart.

I thought of my earlier impressions, when I was fourteen. Then there was wild air, wild distance, and wild soundscape. The landscape was unbroken by human modifications as far as the eye could see, over it was wild air, and within it a wild soundscape of silence alive with natural sounds. It is difficult to gauge the importance of these wild elements. When they are abundant, we take them for granted. They are part of the context. When they are encroached upon in a piecemeal and gradual way, we might not notice their diminution.

As we sat on the rocky summit, I counted jet overflights. Every fifteen or twenty minutes the silence was broken by the long deep rumble of commercial jets. We couldn't see them, but their frequency and noise were distressing. When I remarked on this, my friend said that she had scarcely noticed them and that I would enjoy the place more if I ignored them, much as we ignore pesky mosquitoes. But, I realized, it is quieter in my study in the city than it is now in the heart of this million-acre wilderness. These thoughts drifted through my mind, the planes continued to fly over, and I thought of wild silence, wild air, and wild skies, as

I have experienced them through the decades on the Olympic Peninsula. Over the years I have lived in many places, but the Olympics have remained my spiritual home. No matter where I lived, I always returned to them.

Before I even set foot there, I grew to love them from afar, seeing them through the seasons from twenty-five miles away. I must have been under ten when they first appeared to me, mysterious in distant clouds and mist. Their mystery has ebbed and flowed as I've wandered over the peninsula and come to know them through the seasons. My knowing them has reflected my coming to know myself.

How have these passing years reflected the waxing and waning of the wild and controlled in these sacred places? This waxing and waning, as reflected in ourselves and the land, is especially vivid when we have a long connection with a particular wilderness. Through the practice of wilderness, journeying a whole mountain range becomes internalized. It penetrates the mind and bones. Like the first human inhabitants, we develop a deep connection with place. We become indigenous. The first dwellers identified with their communal group, which was identified with their place. Their memories, their joys and sadness, relatives and ancestors, became immediately identified with the spirit and wisdom of the places they inhabited. They fit into them. In contrast, modern industrial society is placeless. It modified places to fit a preconceived plan, rather than fit itself to the wisdom of a place. I could see this very clearly as I looked out over the peninsula where destruction of place and indigenous wisdom of place are both in evidence.

———

This process of identification is not a projection of ego or small self. It is not a projection of our own dark shadowy side. It is a process of *reciprocity* which reveals the larger ecological Self. If we follow our breathing, as we journey in the mountains, we realize our intricate interconnectedness with everything in wild space. The context is neither created nor is it controlled by any small group of entities. It is mutually co-created through the reciprocal communal activities of myriad beings within the physical realities of the place. We begin to understand the place-specific flow of energy and creative power in the natural world when we see it through complementary, indigenous, human and nonhuman communities. To see in this way our eyes must be free of a desire to rebel or control. Understanding depends on the conditions of our own heart.

Communal dwellers see a different world than rogues. A *rogue element* ignores or rebels against communal constraints, but an *imperial rogue* attempts to control the context by means of centralized plans and force. There are many types of human rogue elements and imperial rogues. Historically some rogue elements had so much energy and charisma they were able to control communities and get tribal energy to identify with their rogue plans. Sometimes a group's size and organization give rise to an imperial rogue (Rome, for example) which attempts to impose its image upon a whole context, even the earth.

Traditional Native cultures on the Olympic Peninsula were not imperial rogues. Their impacts on the land after centuries of habitation were minimal. They did use fire and other natural agents to improve gathering and hunting, but these were local in scale. When the first Europeans arrived, the land showed little noticeable human impact after thousands of years of habitation. Today, after only a few centuries, the violent impact of industrial culture is starkly evident. This impact is one of the clearest examples of the imperial rogue behavior of industrial culture, and it acts the same way everywhere in the world.

When one views this impact from above, one is struck by its geometric structure and the pervasiveness of its damage. Its "forestry," for example, is not a tending or caring for forests; it is forest removal. Forests are erased, everything burned and razed, as in preparing a field to be planted with corn. Nursery trees are planted in the fields, sometimes only one "desired" species. This mechanical control represents an abstract notion of wilderness. Wilderness becomes an abstract concept, a legal category, an interest to be divided, a human invention. For Western industrialism there is no such thing as specific wisdom of place. Rootedness and commitment to the spirit of place are not valued.

Indigenous communities, however, have deep rootedness through time in specific places. They develop a deep ecosophy (ecological wisdom and harmony) that is specific to their area. In the process they contribute to and participate in the creation of the place they inhabit. A place is all sorts of things, including vernacular knowledge and wisdom. But this is not recognized in industrial paradigms, for they neither value nor recognize this process of reciprocal participation with nature.

———

As we sat on the summit in the heart of the Olympics, everywhere we looked beyond the park boundaries were vast clearcuts: large, bare, brown rectangles and squares cut out of the landscape. Those that had greened by natural regeneration looked like they were healing. Some of the planted areas looked uniform, unnatural, unhealthy. They did not fit in. Beyond the formerly old-growth forestland we saw suburbs, and then cities. The gray *smerge* of their pollution was highly visible. Odors drifted through the air, faint but there. To the southwest we could see the giant cooling towers of the unfinished nuclear power plants at Satsop beyond the south boundary of the park. We realized for the first time that the whole park is in their fallout zone. But then, the Chernobyl disaster had already shown that the whole planet has become a fallout zone to the effects of modern industrial technology's violence.

We looked up at the sky. Overhead were contrails of commercial jets carrying the wealthy of the world to do their business, recreation, and government activities. With them flew the same uniform systems of control, communication, finance, monetarism, infrastructure. This is spread worldwide in order to "modernize" everyone. Modernization destroys the character and uniqueness of wild and cultural places, but this is rationalized as necessary for "development and progress."

I compared the feelings, thoughts, and perceptions I had had on this peak over forty-five years earlier. How different my sense of the wild and controlled had been. In the intervening years I became a member of a growing movement working to preserve treasured, sacred wilderness places. From the early years of the movement there has been expansion toward deep questioning which reconsiders our basic values and relationships to nature. The result is a movement toward ecocentrism that diverges from humans-first conservationism. Early in the environmental movement, John Muir and Gifford Pinchot parted company over what seemed specific issues of preservation regarding Hetch Hetchy. But their differences went far deeper.

Muir championed wilderness preservation because he came to identify more and more deeply with wildlands and their intrinsic values— not just with certain "special" places, but with "ordinary" ones. He saw how our ways of thinking and acting can limit our perception of the world-as-it-is. He realized that we cannot isolate ourselves from the ecosphere. What we value determines the technologies and economies we build and how we participate. Even with the best intentions, as in trying to share our "treasures" (modern technology) with others, we end up being global rogue imperialists because we try to bring the whole *context* under our total control.

Pinchot stood for a wise-use philosophy which does not recognize the inherent value of other beings, only their instrumental value to humans. Thus we should use these resources responsibly to create a sustained yield for future generations of humans. This philosophy lends itself to rogue imperialism. It does not question the values of industrial culture with its ideas of development and progress. Muir, in contrast, discovered the intrinsic value of other beings by coming to know them intimately through wilderness wandering.

When we engage in wilderness journeying free of conquistadorial aims, we enter a wild context as pilgrims, as wanderers, as guests who pass through almost unnoticed, not as rogue intruders or imperial rogue conquerors. The spiritual dimension of wilderness journeying opens us to the spontaneous, creative power of nature. What unfolds around and within us is aesthetically perfect. It is not the product of a central plan. It is created through the complex *participation* of multitudes of beings who live in harmony with the places they themselves help to create in community with other beings. This is different from the centralized thinking and tinkering which dominates industrial paradigms and modern scientific rationality. For them "the wild" is part of an abstract theory whose aim is control for human use. Progress means only more extensive and efficient control, which brings more profits and greater pleasure for humans. Controlling the wild is not equivalent to living in harmony with it.

The modern Western idea of progress is built on Christian millenarianism with its otherworldly emphasis on salvation and the Second Coming. The church extended the patriarchal hierarchy of Imperial Rome, and the philosophy and worldview of Christianity became an imperial doctrine used to rationalize the conquest of indigenous, non-European peoples. Elements from the worldview of Christianity, including its philosophy of history, became part of modern scientific rationality with its emphasis on technological development. Central to this approach is the view that there is one superior way to do things, one superior solution to every problem, only one kind of decent future for all humankind, rather than many futures. It pursues uniformity in methods and practical knowledge that can be centralized. When coupled with industrial technology and the money economy it destroys diversity and pluralism. Indeed, it even seeks to patent wild knowledge and genes. Christian philosophies gave industrialism the idea of a universal paradise on earth which appeals to those living in exploited conditions (sin). Technology and science represent the cross and church; Technotopia replaced Paradise. God's creation became raw nature, a storehouse of materials to be processed by industrial technology in the name of progress (salvation), so today we can live in a monetized time. Even wild time is cultured in a world rapidly becoming a technological artifact. Wilderness areas become managed parks, while museum dioramas try to imitate the wild. Even "experience" is manufactured and sold.

It is difficult to appreciate the degree to which the wildness of the Olympic Peninsula has been penetrated by the whole spectrum of industrial culture's "development" activities, ways of thinking, and byproducts. As I looked out over the surrounding wilderness, a land created by spontaneous processes involving countless beings and myriad communal interrelationships, I thought about how thinking in terms of parks and wilderness parks is itself a form of participation in the global anti-wilderness structures of rogue imperialism. We unwittingly miss the larger context, even though we opt for a park of one million acres. It still is not enough! It is a refuge—an island surrounded by violent impacts of an industrial monoculture which is destroying complex, diverse, communal systems that have taken millennia to evolve. In their place it leaves devastation and uniformity, impoverished ecosystems, debt-ridden economies, and damaged humans.

———

As I sat on the summit of the unnamed peak, I reflected on the waxing and waning of my own wild consciousness and its identification with the fortunes of the Olympic Peninsula. I realized, as I surveyed the last half century, that the peninsula is a microcosm of human relationships to one another and to nature, showing both our successes and our failures to attain harmony. Harmony—or lack of it—is one of the central features of relationships. Harmony with another person implies that there are two elements to be harmonized through reciprocal relationships. In the Industrial Market model, by contrast, where the number one, biggest and fastest is the best, the aim is to dominate the other in a relationship. This implies successful control, even over one's own inner ebb and flow of feelings. This also implies that there is only one center in a relationship—a center which imposes its form of order on all others in the relationship. At the other extreme there is the chaos which has no center—no shared context, whether of imposed uniformity or of mutual respect for richness, abundance, and diversity of relationships and values. In natural communities there is a common civility that allows a plurality of centers to relate to one another harmoniously. For humans such harmonious relationships are in part the result of conscious choice. As self-aware moral agents with spiritual dimensions, we live through our choices.

Attempts to control wild beings are usually based on failing to understand who we are. Knowing who we are is a task necessary to wisdom. Wisdom is not the same as knowledge, for it is an unqualified good. Wisdom is like the unconditional love that parents should give to their children. Wisdom arises when we have such unconditional love for our place and ourselves. Knowledge as power is always a qualified value, unless it is knowledge of the good or self-knowledge. Self-knowledge and understanding of the good are bound up together. Who we are requires a deep inquiry into the extent of our self-ignorance and identifications. As a Native American elder once said to a great warrior, "If you do not know that you are these plants, these rocks and mountains, these birds in the sky, this ground under our feet, these animals, you do not truly know who you are."

The technology of Western industrialism relies on technique, cleverness, and control, not on wisdom. It is governed by smart competitiveness and crafty power. In the face of its material abundance, we lack the vital necessities for dignity and self-realization. We are burdened by the things we have made, yet short of the values we should provide to one another, our communities and bioregions. We deceive ourselves about the negative impact our lifestyles have on the natural world, other humans, and other beings. So we do not accept responsibility for examining our lives and the consequences of our choices. We do not see how the wilderness and the land reflect who and what we are as a people. There we can see the results of our values and commitments, the depth of our compassion and wisdom. What must we do to meet our vital needs but minimize our harm? We are part of the natural world, born of the earth. We are not extraterrestrials. What we do to the earth, we do to ourselves. How we act reflects our depth of regard for our larger ecological Self.

I sat on the summit and tried to think of humans as beings from outer space. Western culture behaves as if it is not part of the earthly context. In self-ignorance we act as if we can control the natural world as something apart from ourselves, by means of technology based on scientific knowledge. Progress is defined only as improvement in power, efficiency of control, and profits. Success is measured only in dollars. Only the specialists, the highly talented, and the ambitious really succeed. The rest of us become passive recipients instead of active contributors and sharers. Industry manufactures entertainment and we lose our ability to tell our own stories.

I see how the peninsula and its core wilderness reflect the layers of the human self cultivated within the complicated historical strata of Western industrial culture. Our activities and values are on full display in the recent history of this place: the scars on the land; the psychic damage to families and communities; the wasted lives; the hopelessness; the melodramatic pursuits to avoid self-awareness; the suffering of other beings and the destruction of their homes (habitats). Our denial of death is a shadow cast by our own technology-magnified activities which threaten to kill the wilderness heart. The options we have closed measure the despair hidden in the worlds destroyed by blind participation in lifestyles that are destructive. Perhaps we ourselves have been functionaries, components in a larger technological machine.

The mountains of the peninsula look worn and stressed around the edges, and even at their center there are signs that all is not well. Industrial acid rain comes to the peninsula. It is thought to originate in Japan and South Asia. Who knows? Does it make a difference where it originates? We know what kind of lifestyles and technologies generate it. We know why the salmon and other wild fish are disappearing. We know why the ozone layer is thinning and why the rate of extinction of species is rising. And as technological agents we must accept responsibility, even as conditions grow worse.

If we look at the Olympic Peninsula and its natural inhabitants, its indigenous beings, we can trace the history of our changing perceptions and actions in thousands of different ways. Think of these changing perceptions and actions as complex stories within stories. The smallest stories contain subplots and themes. The more characters a story has, the richer its themes, the more complex its diverse values and possibilities. Here we must be careful to use the nonviolent language of earthly civility. How do these stories reflect an expanding sense of identification with a larger ecological Self? I am trying to answer this question by means of personal reflections on the Olympic Peninsula. Am I still an exotic being naturalized, or have I become an indigenous dweller in this land? I identify most strongly with Southern Vancouver Island and the Northern Olympic Peninsula, but my ecological Self includes the whole island and the whole peninsula.

———

I have felt the ebb and flow of control and its pathology in my own life. I have been controlled by others, have controlled others, and have tried to control the world around me to a degree. I know now that this attempt at control was related to my fear of living through pains that had accumulated over the years. It is a familiar story. We fear to acknowledge our deepest feelings and live through them. I tried not to extend this control to the integrity of the natural wild world. But I did not appreciate the degree to which wild air, wild sky, and wild soundscapes are a context we participate in creating. It was not until I experienced giving up such control that I began to live an authentic spontaneous life. So I learned the obvious: the world we experience is, in part, created by how we react to what happens to us. Angry people perpetuate an angry world. Joyous people find the joy in the world and learn that a simple life can be rich in meaning and deep in happiness.

We must choose wisely. Given the impact of technological society on natural communities, it's clear we've reached a critical juncture. The new strip slurbs of the North Peninsula exemplify the destruction of harmony and natural integrity which result from having technological power and the wrong values. Yet it is possible to see in our society the seeds and shoots of a new culture that is rooted in sustaining a context of respectful civility, one which facilitates harmony among a great diversity of cultures. Diversity of cultures depends directly on preserving biological diversity and restoring natural diversity where our monocultures have destroyed it.

The environmental crisis is a crisis of Western industrial culture, character, and consciousness. The destruction we see in the natural systems and wild world of the peninsula and around the earth is wrought by models of uniformity and control. We must free our minds of these homogenizing structures and find our own inner wilderness core so that we can release our visionary and spiritual powers to transform the way we are in the world in order to create harmonious relationships with each other and nature. We need to envisage a long transition to sustainable communities, a world of natural abundance in which there is a great diversity of human cultures, a plurality of celebrations of intrinsic values we discover as we journey through wildlands—fully conscious, realizing our interconnections, and living in the joyful wisdom of compassion, discovering the wisdom of place.

Today when I walk the ridges and valleys of the peninsula, I feel its many forms of life, its flows of energy, the wisdom of its places, its spiritual powers. I realize that its whole story is the global one writ small. Earlier we came to the mountains to conquer the summits, but today we humbly descend. For when we reach the top we look out to mountain ecosystems which have limits, but within are summits and valleys that have no end of richness.

I used to think that wilderness complemented the city, but now I realize that wilderness *completes* the city. Without wild forests and mountains, in the absence of wilderness, all cities would cease to exist. The sustainability of a city is directly related to how well it harmonizes with its place. It needs to be seen within a larger wilderness context as well as to hold the wild within it. This is true for a community; it is true for a culture; it is true for the self.

Human cultural activity writes its story upon the landscape. Where we minimize that impact, we leave open the greatest possibilities for self-rediscovery. We journey to the mountains to find what's there, and we find ourselves in their mystery, reflected in the layers of history and time, reflected in the daily impacts and in our ongoing relationships. We come to know the ecological Self when we forget the small, historical ego. The ecological Self is the transpersonal self; it is a context created through reciprocity.

I recently attended a birthday celebration for a friend who is a Native American shaman. It was held at his home on ancestral lands near Victoria, B.C. He has lived there for fifty years, as his ancestors and his people have lived for thousands of years. His identity is interwoven with the place in which he resides. He is not a visitor. He realizes the power and wisdom of the place. His house looks out across the Strait of Juan de Fuca to the Olympic Mountains. During the party he performed a ceremony in his own tongue, translating for us as he went along. While he was talking the whole area seemed to grow still. I could hear the birds far across the bay as they answered his echoing voice. He, his culture, his lineage, his place, nature and its spirits, all were in harmony.

What has happened to the Olympic Peninsula since the coming of Western culture reflects what industrial society is doing worldwide. We can rationalize this imperial rogue behavior no longer. We must change our culture, settle in place, and become authentic indigenous natives who dwell in the land. Our modest contributions to harmony will be small repayments for our rogue behavior and tacit support for rogue imperialism. In wilderness and the wild is the salvation of our selves. May all wild beings flourish and realize themselves!

Sacred Geography of the Pikuni: The Badger–Two Medicine Wildlands

JAY HANSFORD C. VEST

In a cautionary note to the planners of cities, Plato declared that certain locations possess ecological and spiritual qualities which markedly affect development of human character. This advice echoes through time with an essential wisdom for contemporary land-use planners. The area known as the Badger–Two Medicine, located along the east slopes of the Continental Divide just south of Glacier National Park, is a place that deserves the sagacious consideration urged by Plato.

These wildlands were given reservation status through an 1855 treaty between the United States and the Pikuni (Blackfeet) Indians. However, an 1896 agreement between the two parties ceded the Badger–Two Medicine to the United States; the Pikuni reserved the rights "to go upon" these lands, to hunt and fish there, and to gather the area's timber for personal use. The Badger–Two Medicine later became part of the Lewis and Clark National Forest.

Traditional Pikuni respect the Badger–Two Medicine as sacred land. According to religious scholar Åke Hultkrantz, it is a region of sacred geography wherein "nature is sacred because it reveals, or symbolizes, the Great Mystery." Thus any disturbance of the area's natural integrity desecrates its power and diminishes the central values of tra-

This is a revised version of an essay that originally appeared in *Western Wildlands* (Fall 1989).

ditional Pikuni religion. It is precisely in this context that controversy arose during the forest planning process.

Although the Badger–Two Medicine is a de facto wilderness, strategically located between Glacier National Park and the Bob Marshall Wilderness complex, the U.S. Forest Service leased the lands there for oil and gas exploration in the early 1980s. Forest Service planners declared that "the area is not available for wilderness classification because of rights retained by the Blackfeet Tribe in the Agreement of 1896." The agency further argued that under the agreement "the Blackfeet Tribe retains the right to cut and remove timber," so the area cannot be designated wilderness (U.S. Forest Service 1986). Not only is this defense incorrect, but it neatly sidesteps the fact that the earlier decision to allow oil and gas exploration already interferes with Pikuni rights to practice their traditional religion in the Badger–Two Medicine.

Pikuni traditionalists joined conservationists in appealing the Lewis and Clark Forest plan, which would facilitate a comprehensive program of oil and gas development. Their objection to development in the Badger–Two Medicine is hardly frivolous. The area represents considerable spiritual value in their religion, which can best be understood through examination of the role of the vision quest in traditional Pikuni religious practice.

Religious scholar Joseph E. Brown considers the vision quest to be the foundation of Native American religions. It is, he explains, "the most profound spiritual dimension at the heart of these cultures." While in retreat, a person on a vision quest "opened himself in the most direct manner to contact with the spiritual essences of the manifest world. . . . No person in these [Plains Indian] societies, it was believed, could have success in any of the activities of the culture without the special spiritual power received through the quest."

Among contemporary Pikuni traditionalists, visions remain central to the process of receiving religious power. For both Pikuni individuals and the tribe itself, knowledge of the future is obtained in the solitude of "a remote region among the Rocky Mountains" (Grinnell 1962). The essential nature of this retreat into wilderness solitude is exemplified by the Pikuni belief that dreaming at home may be abortive and fail to convey power (Wissler 1912). Hultkrantz confirms how essential wilderness is to vision quests: "Typically, the vision of the guardian spirit is individual, sought and obtained in solitude and isolation out in the wilderness—for example, on secluded mountains and hills."

The Pikuni religion has a rich tradition of myth that encompasses the Badger–Two Medicine area; indeed, myth is central to the religion. Mythic themes express sacred events in the now: they are "time outside of time." Myth informs and explains reality—not in the linear fashion of science but in the events of ever-active creation, occurring and recurring in the cycles of days or seasons and in death and rebirth.

The Pikuni call the Rocky Mountains the "Backbone of the World." A sacred landscape, the Backbone is home to "powers," "spirits," or "other than human persons," including Thunder, Cold Maker, Wind Maker, the Medicine Elk, the Medicine Wolf, and the Medicine Grizzly among others. Na'pi (Old Man), the Pikuni creator and trickster, is a central figure in many of these myths. The following abridged myth documents the "powers" and the creator's relationship with the Badger–Two Medicine wildlands, hence confirming the region's sacredness.

———

In the long ago, when Na'pi created the world, he separated men and women into two camps. The women lived in the mellow Cutbank Valley, and the men lived in the mountains along the Two Medicine River. Realizing the error of this separation, Na'pi said to the men, "You shall no longer live by yourselves. Come. We will go to the camp of the women."

All of the men were most willing to meet the women, so they dressed in their finery, particularly Na'pi, who was the finest looking of all the men. At the camp of the women, the men all stood in a row because the women had the right of choice. Although she was poorly dressed and dirty from butchering buffalo, the chief of the women had first choice. She walked up and down the line of men and finally returned to Na'pi, taking his hand. Na'pi noticed many fine women waiting their opportunity to select a mate; intrigued by the others, he rejected the Chief

Woman. Angered by his rejection, she returned to her camp, cleansed herself, and dressed in her finery.

Returning to the hill where the men stood, the Chief Woman appeared transformed; indeed, Na'pi thought her the best looking of all the women, and he kept stepping in front of her so that he might be chosen. Ignoring him, the Chief Woman chose another man, and all the other women followed her and left Na'pi alone. Na'pi became very angry that he was not chosen; because of his behavior, the Chief Woman turned him into a pine tree, which stands alone at the edge of the mountains where the plains begin.

It is said that Pikuni romance endures today because of the beauty of the site of this first mating. The editors of *Historic Montana* reported that Pikuni men lived in the mountains south of the Two Medicine, which means they were in the Badger–Two Medicine wildlands. And as "Keepers," the Chief Woman (keeper of women) and Na'pi (keeper of men) require an undesecrated place.

———

McClintock (1968) found that among the Pikuni, "the Great Spirit, or Great Mystery, or Good Power, is everywhere and in everything—mountains, plains, winds, waters, trees, birds, animals." As this power is an endowment from Natos (sun), which is acknowledged as the creative source of all power and animation, Natos is venerated by the Pikuni. In turn, Natos gives his blessing unto the people "for their reverence for all of nature." The reasons for praying to Natos are explained in the story of Poia (Scarface), the most ancient tradition of the Pikuni religion (McClintock 1968). On a warm, cloudless night, Soatsaki (Feather Woman) was sleeping outside her tipi in the long grass when Morning Star, rising beautifully above the prairie, came and made love to her. She found herself with child, yet she was a pure maiden, for none of the men had been with her. Coming to earth and making himself known to her, Morning Star invited Soatsaki to his home in the sky, where she went to live with him and his parents—father Sun and mother Moon.

In time, Soatsaki gave birth to a child called Star Boy. Soatsaki and

Star Boy were banished to the earth when she disobeyed the sun's command and dug a large, sacred turnip. Going each morning before daybreak with Star Boy to the summit of a high ridge, Soatsaki mourned her banishment and pleaded with Morning Star to take her back. In her grief, Soatsaki died, and Star Boy was left alone without relatives. He was subject to much abuse because he was born with a mysterious scar on his face; in derision, the people called him Poia (Scarface). Poia was rejected by the maiden he loved and then learned that only the sun, Natos, could remove the scar.

With the help of a kindly medicine woman, Poia journeyed to the mountains, following the path of the sun. When he reached the sun's lodge, Natos agreed to remove the scar. In doing so, he appointed Poia to be his "messenger to the Blackfeet, promising that if they would give an Okan (Sundance) in his honor once a year, he would restore their sick to health. He taught Poia the secrets of the Sundance and instructed him in the prayers and songs to be used.

Poia then returned to earth and the Pikuni camp by the Wolf Trail (Milky Way), and there he instructed his people in the ways of the Sundance. Subsequently, Natos took him and the girl he loved to the sky, and there they became a bright star—Mistake Morning Star (Jupiter).

During his quest, Poia traveled first to the Sweet Pine (Grass) Mountains. There he was told to "go far to the west to some very large mountains and to the highest of them all. Scarface must sleep there to seek out the spirit of the mountain" (Bullchild 1985). In traveling west from the Sweet Pine Mountains, the highest peak encountered is Morning Star Mountain (8,376 feet), part of the Badger–Two Medicine, which also encompasses Feather Woman Mountain, Scarface Mountain, and Mount Poia.

The names of these peaks are derived from the central Pikuni myth and are seen as part of a mythic landscape—thereby the most sacred of lands. In the traditional naming of a thing, the name imparts the essence of its meaning upon that which is named. Since the essential meaning given here is of the most sacred character, and because it is believed that the myth lives with the names, the landscape given these names is assuredly sacred.

———

The grizzly bear has a sacred role among traditional Pikuni. The animal is one of the most powerful totems, or helpers. Because of these powers, compounded by the bear's endangered status, the Badger–Two Medicine wildlands are of great importance to traditional Pikuni religion as home to these sacred animals. In the legend of the friendly Medicine Grizzly, the Chief Bear befriends Blackfeet culture hero Nistae (Calf Robe), who has been wounded in a raid. The Medicine Grizzly feeds him and carries him northward along the Backbone to the Pikuni camp at Bear (Marias) River. In gratitude, Nistae invites the Medicine Grizzly to live with him. The bear refuses, however, saying: "The moon is now near past when the leaves fall off. It is time I should find a den, for the heavy snows of winter will soon come. The only favor I ask of you in return is that you will never kill a bear that has holed itself up for winter." Turning westward to the mountains, the Medicine Grizzly departs for his home. As a result of his request, the Pikuni will not kill a hibernating bear. This myth establishes a fundamental moral relationship between the Pikuni and the Medicine Grizzly, who represents all "real bears." It is significant that the Medicine Grizzly went west to the mountains from the Pikuni camp. Following the primary fork of the Bear (Marias) River, he would eventually have entered the mountains of the Badger–Two Medicine wildlands.

In another account, a young mother, Itsapichkaupe (Sits-by-the-Door), is captured by the Crows and taken more than two hundred miles to a camp on the Elk (Yellowstone) River. There she is pitied by a kindly Crow woman who helps her escape. During her long journey home, Itsapichkaupe's provisions are soon exhausted and she finds herself deep in despair when a large wolf approaches her. As the wolf lays at her feet, however, she beseeches his aid: "Pity me, brother wolf! I am so weak for food that I must soon die. I pray for the sake of my young children that you will help me." The wolf then draws near to her, and Itsapichkaupe is able to walk by placing her hands on his back. Thanks to the friendly wolf, Itsapichkaupe safely reaches the Pikuni camp along the Bear River. The faithful wolf retreats from the camp to the nearby mountains but comes every day to a hill overlooking Itsapichkaupe's lodge.

Believing the wolf and the coyote to be good medicine, the Pikuni never shoot them. Indeed, they have a saying: "The gun that shoots at a wolf or a coyote will never again shoot straight" (McClintock 1968). Like the Medicine Grizzly, the Medicine Wolf makes his home in the Badger–Two Medicine wildlands; again like the "real bear," the wolf is protected in the moral philosophy of traditional Pikuni.

In these mythic accounts, the Badger–Two Medicine wildlands are inextricably aligned with traditional Pikuni cultural religious identity—the myths confirm a sacred geography. The Badger–Two Medicine represents Pikuni traditional sovereignty and recalls a cherished way of life. As these wildlands offer the tribe's alienated and lost a way back into the traditional culture, the area provides a means of recovering a sense of pride and honor in being Pikuni.

Spiritually, the Badger–Two Medicine is a source for the gathering of traditional Pikuni "medicine power" and thus has a significant role in reaffirming the moral fabric of the Pikuni Nation. Accordingly, these wildlands are not only symbolically important, they are essential to the recovery of traditional Pikuni culture from decades of oppression. No monetary settlement can match the gifts of dignity and tribal identity that the area holds for traditional Pikuni culture. Consequently, if the tribe's traditional identity is to be recovered and retained, it must include the preservation of the Badger–Two Medicine wildlands. The Pikuni traditional religion is alive and flourishing today. And it is practiced in the Badger–Two Medicine as a sacred landscape.

References

Brown, J. E. "Modes of Contemplation Through Action: North American Indians." *Main Currents in Modern Thought* 30 (1973):62.

———. *The Spiritual Legacy of the American Indian*. New York: Crossroads Press, 1982.

Bullchild, P. *The Sun Came Down: The History of the World as My Blackfeet Elders Told It*. San Francisco: Harper & Row, 1985.

Grinnell, G. B. *Blackfoot Lodge Tales: The Story of a Prairie People*. Lincoln: University of Nebraska Press, 1962.

Historic Montana. Montana Heritage Series no. 9. Helena: Helena Historical Society of Montana Press, 1959.

Hultkrantz, Å. *The Religions of American Indians*. Berkeley: University of California Press, 1970.

──────. *Belief and Worship in Native North America*. Syracuse: Syracuse University Press, 1981.

McClintock, W. *The Old North Trail: Life, Legends and Religion of the Blackfeet Indians*. Lincoln: Bison Books, University of Nebraska Press, 1968.

Schultz, J. W. *Blackfeet Tales of Glacier National Park*. Boston: Houghton Mifflin, 1916.

──────. *Why Gone Those Times?: Blackfoot Tales*. Norman: University of Oklahoma Press, 1974.

U.S. Forest Service. *Lewis and Clark National Forest Plan—Record of Decision*. Great Falls, Mont.: Forest Supervisor's Office, Lewis and Clark National Forest, 1986.

Wissler, C. "Ceremonial Bundles of the Blackfoot Indians." *Anthropological Papers of the New York American Museum of Natural History* 7 (1912):104–105.

The Land of Absence

MOLLIE YONEKO MATTESON

With bison gone, the last white wolves turned secret to the mountains,
barren in their life of flight. Teeth worn down, his lover dead, we shot the
last one over cattle killed in need. Pretend he isn't stuffed and standing
in a courthouse with the laws that mark your bones in ordered rows.
Burn to ash on wind. Scream across the plain.

MATTHEW HANSEN, *Tribute*

I decided to see the White Wolf one February morning. Cold, dry air snagged my breath as I carried my things out to the car. Notebook, handbag, map, water bottle, peanut butter and jam sandwich, an apple, a sleeping bag. Hoarfrost furred the trees. I scraped ice from the car windows.

"Are you going just so you can look at a stuffed wolf?" my husband asked.

"No," I said.

I paused. "I'll see you tonight, early evening."

"All right." He gave me a hug.

I drove east, then north. From Livingston, Montana, on the Yellowstone River, to Stanford, in the Judith Basin, I figured was a three- or four-hour drive. I was supposed to be writing about wolf recovery, and on the first clear day in a week, I was headed for a place where I supposed wolves would never roam again.

Captain William Clark of the Lewis and Clark expedition, traveling up the Missouri in 1805, named the tributary Judith River for his bride-to-be, then waiting back in Virginia. In the 1870s, the luxurious grasslands of the Judith enticed land-hungry stockmen. The Milwaukee

Railroad made the basin known throughout the nation as it tried to lure homesteaders westward early in this century.

And on May 8, 1930, the last "renegade" wolf in Montana was killed in the Judith Basin.

As I drove out of the Yellowstone drainage and into the watershed of the Missouri, the land rippled and bulged. Streams flowed east and northeast, the road lulled me with its rhythm of syncline and anticline. Along creek bottoms, naked shrubs—willow and red osier dogwood—were lusty pink and rose. Then the road would sweep me up to ascetic benchlands. Sky blared. Snow-caked tawny grass lay wind-welded on leeward slopes.

I crossed the headwaters of the Judith at Judith Gap. Two livestock trucks passed in the opposite lane. A bus followed. Its signboard read "Cutbank Wolves." This was the modern Judith Basin. Cattle drives by semi, wolves in chartered buses. Later I passed the Conrad Cowboys, Choteau Bulldogs, and Browning Indians. Must be a basketball tournament, I thought.

Like the Grizzlies of my alma mater, the University of Montana, the reality of these mascots (with the exception, perhaps, of the Bulldogs) had long ago diverged from the mythological ideal. Cowboys rode Fords and Chevys and worked government forms as much as they worked cattle. Indians rode Fords and Chevys too. These days they talked about basketball, not buffalo hunts. Only the wolves and the grizzlies had not changed much. They were simply gone.

The Judith Basin, unlike the empty steppes of the Sweetgrass and the upper Musselshell drainages through which I had just passed, manifested considerable human investment. Fences were markedly more numerous, and the spaces they fenced in were smaller. Barns, ranch houses, tiny hamlets clustered at crossroads, poked above the blue distance like ducks scattered on a wide lake. I passed cows, sheep, and gravel roads cutting off to the far corners of the basin.

But a landscape that had initially appeared inhabited, tended, and nurtured was mostly an illusion. I stopped at a farmstead perched on a high bench. The house slouched like an old woman, its roof sloughing groundward. The unpainted, weathered clapboard was peeling and cracked like badly chapped lips. The wind blew at will through shattered windows.

Twenty miles outside of Stanford, the town of Moccasin could have modeled the effects of the neutron bomb. The empty carapace of a bank, a silent school, flapping doors, eyeless house fronts. Most of those tiny silhouettes on the horizon, it turned out, marked not prosperity but abandoned ambition.

This is not to say that the Judith Basin was a desolate wasteland. Even on that cold day, Highway 200 between Great Falls and Lewiston fairly bustled (by central Montana standards) with traffic: mothers in mini-vans, young men packed three abreast in the cabs of Broncos and Rams, semis towing frozen foods, a bread truck. Every few miles I passed scattered herds of cattle.

Isolated mountain ranges—the Little Belts, the Judith Mountains, the Moccasins, the Big Snowies, the Highwoods—surround the basin like a great inland archipelago. A modest orographic effect funnels additional moisture into the prairie basin, making it "well watered" by standards of the high plains.

Thus more people came to the Judith Basin than tried the upper Musselshell or other, more blatantly arid, places. There were more towns here because the settlers had stayed long enough to build real communities. But the Judith was going the same as the rest of the depopulating Great Plains. The same patterns: ranches getting bigger as ranchers themselves got scarcer; small towns vanishing; bigger towns growing. The farmers and ranchers of the Judith Basin were not going to escape the fate of their peers. They were just going to hold out a little longer.

I arrived in Stanford about lunchtime. Though it was a weekday, the downtown was quiet. A sign caught my eye: Wolves Den Cafe. A wolf's portrait was painted on the front window. Inside, two men in work clothes—creased jeans, baseball caps, flannel shirts—hunched over burgers and fries. Two women held a tête-à-tête in a corner, picking at each other's food. I sat up against the wall, beneath a photo of a contented-looking, restful gray wolf. I ate potato soup and a BLT.

What did the women running the café—the young waitress, the older one who occasionally peered out from the doorway to the kitchen—think about wolves, I wondered? Or *did* they think about wolves? In Vermont, where I grew up, the name "catamount" was commonly applied to businesses, sports teams, and city streets. The last eastern panther had been killed in Vermont in 1881, but that was irrelevant. The image was part of a cultural landscape, and while memories of the living animal faded with the generations, the symbol survived.

I spooned up the last of the soup and ordered a cup of coffee so I could legitimately loiter for a few more minutes. Glancing out at the sunny, still street, I thought about the day over fifty years ago when the White Wolf's killers had paraded his carcass through town. The street had been mobbed. So thick was the crowd of onlookers that the car carrying the dead wolf took several hours to drive the two blocks along Main Street. Sometime after the excitement of that day, he'd been installed in the imposing Judith County Courthouse. It stood at the head of Main Street, its somber facade overlooking a nearly treeless town.

The other customers left. The waitress bustled from one table to another, replacing sugar packets, shoving salt and pepper shakers back together in tidy couplings. I cradled the white coffee cup between both hands. The light seemed very bright, pressing through the clean, large windows. I got up to pay my bill and headed toward the courthouse.

———

In their journals, Lewis and Clark reported a great abundance of wildlife on the Montana plains, including wolves. Accurate estimates of the pre-white settlement population of wolves appear to be nonexistent. I have seen figures such as "in 1800 [wolves] exceed 350,000 individuals" and not known whether to gasp or laugh. One historian suggests that trappers and bounty hunters took 100,000 wolves per year between 1870 and 1877. Today in all of Canada, where wolves still occupy about 85 percent of their former range, the wolf population is put at 50,000 to 65,000.

I give up trying to rationally evaluate the numbers. It is enough to know there were many wolves, more probably than I am capable of imagining.

The advent of fur trappers in the 1830s had little direct effect on wolves, except when a trapper surprised one raiding his food cache or beaver trap. It was not until the 1850s, when the beaver were trapped out, that wolf pelts became valuable. Fur traders sought buffalo hides primarily, but the wolves that followed their carcass-strewn paths were easy targets. By the mid-1860s, the American Fur Company at Fort Benton on the Missouri was taking between 5,000 and 10,000 wolf pelts a year.

The earliest "wolfers" (the men who made a regular occupation of killing wolves) were seasonal. During the summer, they'd work on steamboats, in the mines, or, by the late 1870s and 1880s, on horseback herding cattle. In winter, they'd stock up on strychnine and ammunition and lace the country with poisoned carrion. Entire wolf packs, sometimes several packs, died at these bait stations.

By the 1870s, stockmen were moving cattle onto the public domain lands of the high plains. The hidehunters had decimated the once-vast buffalo herds. White settlers and the U.S. government would complete the subjugation of the Indians within two decades. As for the wolves, they turned to eating livestock. Where wolves had been merely pests to the mountain men and later, to the early wolfers, a means of picking up extra cash, they became a direct threat to stockgrowers.

The first bounty law was passed in Montana in 1884. One dead wolf, one dollar. In addition to strychnine, the wolfers and ranchers employed other techniques: traps, snares, and packs of dogs. In spring, wolfers would locate a den, hook pups out one by one with a piece of wire, and club them to death. Some chose not to waste their time with this tedious method, and instead just tossed a stick of dynamite down the den hole. In addition to the state bounty, various stockgrower associations offered their own bounties, for wolves as well as bears, mountain lions, and coyotes.

Bounty prices moved up and down with the temperament of the legislature. At one point, legislators, backed by powerful mining companies, attempted to abolish the wolf bounty. It was a cost-cutting measure that ranchers did not appreciate, and they rallied to have it reinstated. By the late 1890s, the issue seemed moot. Wolves had become quite rare and bounty prices had been declining along with the bounty's practical necessity.

But for some, this end was not final enough. In the early part of the twentieth century, stockgrowers pushed to raise bounty prices again, increasing the price for wolf pups to $5 each. By 1911, the bounty had reached its peak at $15. While ranchers in the more mountainous regions of the state were by that time claiming the complete eradication of the wolf, livestock growers on the plains continued to clamor for "predator control."

Perhaps the reason wolves held out longer on the plains is connected, in some way, to the reason they were so numerous there originally. Not the cold, rugged peaks of the Rocky Mountains but the grassy rangelands hosted the highest densities of native prey species. In the early twentieth century, game was scarce everywhere in Montana—wiped out by overhunting—and the easiest, perhaps the only, kind of living left to a wolf was preying on livestock. In the central portion of the state, the immense openness of the plains themselves offered a kind of hiding cover. The scattered, timbered mountains dotting that sea of space offered another. With cattle plentiful in that country, it was a strange and yet logical place for the last wolves to hold out.

Inside the Judith County Courthouse, a flight of steps led up to a rotunda. Floor and ceiling were pierced with a large circular opening, giving way to views of stories above and below. I half-expected a glass dome above, a sun-speckled hall with potted palms below. But I looked up to a plain, flat ceiling and down to a barren, dark space with fake-marble floor tiles like those seen in elementary schools. A couple of metal folding chairs stood against one wall.

A large mural portraying a sweeping landscape and people with straining, hopeful faces covered the back wall of the rotunda, but I couldn't make it out well. Ladders and scaffolding obscured it.

The White Wolf, in his glass case, was shoved up against one side of the hall. No one entering the building could miss him: his fierce countenance, his lean animal body, demanded attention. But I'd expected a more prominent positioning nonetheless—at the top of the staircase, perhaps, or in the center of the hallway. I glanced into the rooms on either side of the foyer. A serious-looking man pored over piles of paper in the county appraiser's office. A woman leaned over a desk in the other room. Neither paid any attention to me. The White Wolf had become just another piece of furniture, and an awkward one at that.

He was tall and long, but the brittle pelt seemed to have been draped directly over bones. A newspaper account of his capture and death was posted on top of the glass case. It said the White Wolf was eternally "snarling." That was what some liked to imagine, at any rate. His face was twisted into a grimace; his ears were shoved back, conveying ferocity and cowardliness. But whose ferocity and cowardliness? The bullet hole in the left side of his muzzle was not camouflaged, but simply filled in with putty. It did not seem possible that this could have been the mortal wound. It looked more like a kiss.

———

The White Wolf had been preceded by other well-known outlaws in the state. There was the Three-Legged Scoundrel of the Tongue River Valley, in southeastern Montana. He was killed in 1920. The famous Custer Wolf worked the ranges from eastern Montana to the western Dakotas, and was shot in 1921. The renegade pair, Snowdrift and Lady Snowdrift, claimed the Highwoods at the northwest end of the Judith Basin. While their free-roaming existence was condemned, two of their pups, Lady Silver and Trixie, were stolen from the den and taken to Hollywood, where they were trained and filmed in several movies. A sibling, Lobo, was made mascot of prizefighter Jack Dempsey while he was training for the world heavyweight championship in Shelby, Montana. A government trapper killed the adult female and male within weeks of each other in 1923.

The White Wolf, who supposedly brought down over $35,000 worth of cattle, sheep, and horses during his career, was bountied at $400 by Judith Basin stockgrowers. During the late 1920s his fame grew as articles about his campaign of destruction filled papers around the country. Letters and telegrams poured into Stanford: men wanting to come out and join the hunt, women warning to keep the "children near the house until somebody kills that terrible wolf."

He was finally tracked down when two ranchers spotted him in the Little Belt Mountains. They gave chase with their pack of hounds, cornering him after a couple of hours. Al Close shot the wolf in the head, stating matter-of-factly in an article published soon after, "and that's all there was to it." Twenty-seven years later, he recounted for the *Great Falls Tribune* the same incident, but now he was more philosophical:

And do you know, I almost didn't shoot. It was the hardest thing I think I ever did. There was a perfect shot, the grandest old devil. . . . I thought swiftly that these were the hills over which he had hunted. I knew that it was the cruel nature of the wilderness—the fight for the survival of the fittest—that made him the ferocious hunter that he was. I thought of all the men who hunted him, of how his fame had gone out all over the country, and I *almost* didn't shoot. Swiftly these things passed through my mind as I stood there with my rifle aimed, finger on the trigger, and luckily I came to my senses in time and let the bullet fly fairly into the face of the old criminal.

Another account of the day's events described the eager but respectful crowd that pushed into Stanford's main street that afternoon for a glimpse of the animal. "Nobody cursed him," wrote a reporter from the Spokane paper, the *Spokesman Review*. "In spite of the cattle and sheep he had destroyed, everybody respected him. He had possessed a cunning and strength far beyond that of any wolf. He was, without a doubt, the largest wolf ever taken in Montana, and possibly the largest in the Northwest."

And thus it is that we build up our enemies in their defeat, as evidence of our own courage and the fairness of the fight.

———

In the 1920s and 1930s, as the state's wolf population dwindled to almost nothing and the threat wolves posed to the livestock industry faded to insignificance, the individual animals themselves took on the aura of the supernatural. Author and Stanford librarian Elva Wineman, who wrote extensively about the White Wolf, described him as

"the flying gray wraith" who struck "terror and death into the herd and
. . . [fed] like an epicure on the choicest animal of the lot." He was an
"agile spirit," the "monarch of the wilds," a "cunning strategist," the
"mysterious gray-white essence of Satan." Ranchers came to view their
relationships with such animals, antagonistic as they were, as personal.
Montana historian Dave Walter writes: "The killing of such a wolf was
inevitable, but in the meantime he became somewhat of a local hero."

I am stuck on that phrase, "the killing. . . was inevitable." The feeling held by the men who pursued the last wolves was not love, but something that could seem strangely like it: respect; awe; a desire to be close
to the subject of their obsession. And yet the only end they could envision, that they could act out, was one of annihilation.

That's where my comprehension runs out. Is it a cultural barrier? Is
it one of those irreconcilable differences between a man's way of seeing
the world and a woman's? Or could it be the gap between generations—
the difference between growing up in a time when the world seemed big
and people small, and now, when the world is a tiny, limited, tenuous
thing and people seem to be everywhere, seem to be *in* everything?

Couldn't Al Close have stopped at that last second? Why did he
brush aside those stirrings of compassion he purportedly felt? A society
that must carry out its campaign of domination to its ultimate, extreme,
ugly end—take no prisoners, kill the women and children, too—is in
great jeopardy of annihilating itself. The bitter hollowness of that last
act hangs on. It fixes people in a perpetual attitude of defense, forces
them to justify long after even they have lost their righteous certainty.

———

Thus my pessimism about the restoration of the wolf and the native ecosystems of the Great Plains. I believe it will come—that there will be
large prairie preserves, national parks, that even the private lands will
be given over increasingly to all forms of wildlife. But we need to imagine it first. And that will be hard, because the settling of the plains was
so much about, literally and mentally, erasing the last vestiges of
wildness.

In the mountains, deserts, and canyon country of the West, it was far
more difficult to appropriate entire landscapes. The sunny, even face of

the plains was ostensibly more hospitable to early white settlers, and though decades of drought and topsoil erosion have proved that much of the region is not hospitable, the land did submit more readily at the outset.

The poets and writers who sing the praises of the plains, and they are few, do not speak of the wholly wild. They love the people and the landscape—the ranch house tucked in a coulee, the profile of a farmer against the wide sky. And well they should love these things, for they do have their charms. But the darker side is absence. Wolves gone for fifty years. No black-footed ferrets. Two percent of the presettlement population of prairie dogs. No free-roaming bison. Antelope numbers reduced 98 percent throughout the entire West. The willow bottoms long empty of grizzly bears.

Reasonable visionaries, like geographers Frank and Deborah Popper, propose to revive the region's laggard economy, along with the faltering spirit of its communities, by installing a "Buffalo Commons." This would be a mixture of public and private lands geared toward wildlife restoration and management, recreation, and education. A similar proposal, first put forth by Montanan Bob Scott, was termed "The Big Open." While these ideas laudably center on ways to make a naturally functioning ecosystem economically valuable, the plains deserve something more.

The Great Plains must have its seers, its heretics, just as the mountains and the deserts have had theirs. We need a vision of the land not as resource, or even as spiritual haven, but as an entity unto itself: whole, alive, a land of absence giving way to a vital presence.

———

The wolves, for their part, are trying. They have been showing up in eastern North Dakota, single dispersers from Minnesota or Canada. But recovery will not happen, even with wolves appearing voluntarily, until we give them back some of the space we've appropriated. And that will not happen until there is a constituency for wolf recovery on the plains. On private lands, native prey is scarce or intermingled with livestock. Landowners are not apt to tolerate wolves, in any case. Public lands on the plains are limited and scattered, and since these are either surrounded by agricultural operations or are themselves leased out as pasturage for privately owned livestock, land management agencies do not even entertain notions of predator restoration.

In 1991, a government-hired gunner shooting coyotes from an airplane mistakenly killed a wolf. It was an honest error (though one can question the basic value of his mission), because in eastern North Dakota who would have expected a wolf? He did not see the wolf as such because to him the possibility of a wolf did not exist.

We are blinded not by the failure of our eyes but by that of our minds and hearts.

Notes from an
Interrupted Journal

JOHN HAINES

November, 1950. Arrived. All is well. A hard trip from home.
Dogs and I are tired out. It took all day, from daylight to dark.
DIARY OF F. CAMPBELL

July 3, 1980, 6 P.M. We are sitting on a dry hummock of moss near the summit of Buckeye Dome, three thousand feet in a range of hills that stretches east from Fairbanks toward the Canadian border. As far as we can see, looking north, east, and west, the rounded shoulders of the domes, the forested ridges, and the deep valleys unfold, bathed in the warm, clear light of early evening.

We take in this view, this incredible expanse of light and wooded earth. Now that we have arrived, we share once again in the strange exhilaration that seems so much a part of the earth's high places: to be able to look so far and see neither city nor settled land, to hear nothing but the wind over the grassy, sunlit ground.

But there is something else to be seen not far from here, not so grand, but given the times in which we live, not surprising. At the bottom of the steep north slope of this dome, in a saddle formed by the heads of two small creeks named Minton and Rosa, there is a rust-colored streak of cleared ground. It is a small but visible section of the Alyeska Pipeline, cut through this backcountry in the 1970s. With a little squinting from this distance, it might be taken for a highway, but it is strangely silent and without moving traffic. I see it now

This is a revised version of an essay originally published in *The Ohio Review* 47 (1988).

with neither anger nor resentment, but with a kind of wonder that in my lifetime such a mysterious and drastic change could have come about.

———

I have written: July 1980. But I could just as well have written: August 1954; June 1959; or September 1962. My memory of this dome and its surrounding country goes back over thirty years; further back than that, if I include what I know of it from the recalled memory of older and vanished neighbors. I can hardly tell now where my own experience and that inherited from others breaks and connects. I am thinking of a hot day in mid-August many years ago, when Peg and I and Fred Campbell stopped here with seven dogs and loaded packs on our way to a small lake hidden now from our view by a long ridge to the northeast. Hot, tired, and thirsty, we were sprawled near this same spot at high noon, resting before the long plunge downslope, the climb that would take us up that farther ridge, and on to the lake we would come to late in the day. That summer the shrubs on the tundra around us were ripe with heavy blueberries; we easily picked a quart, and ate them with canned milk sweetened with a little sugar from a jar that Campbell kept here on the dome, hidden in a shallow hole in the brush.

In late September of that year Campbell and I came back to the dome with packs and rifles, hoping to find a fat young bear feeding in the meadows. We saw nothing that day, but once while we stopped to rest and to glass the frost-browned, sunlit hillsides, we heard far down in Minton Creek the nasal snore of a moose in rut, like a great fly buzzing.

And once again, early in June of the year following Campbell's death, we came over the dome under windy, gray skies on our way to the lake. We found the lake cabin in poor condition: a bear had broken into it, and nearly everything of value—foodstuffs, bedding, and clothing—was torn and crushed, and there was daylight showing through the sod roof. We rescued from the sodden floor a few pages of a diary that Campbell had kept over the years, written down in pencil on the backs of labels taken from tins of evaporated milk. The next day we made the fourteen-mile hike back to Richardson in a cool, wet wind without stopping.

———

Roads and pipelines notwithstanding, it is still wild country here. From this round and open hill the wilderness seems to have no end. It will be increasingly rare now on earth to be able to look so far and see only range beyond range, valley after valley, unclaimed and unoccupied. And yet at this moment there is about this land an almost pastoral peace. The higher mountain ranges have their grandeur, snows, and glaciers, but the prospect here pleases me more with its gentleness, with its spacious human dimension. And then I think that for so many years this country was the preserve of one man who hunted, trapped, and prospected its creeks and ridges alone until the years compelled him to quit.

From this dome a trail that Fred Campbell kept open for forty years drops steeply to the saddle below us and climbs that farther ridge we see in the middle distance. And the trail goes on, with many forks and byways, all the way to Campbell's Lake, and beyond that to Mud Creek, McCoy, and Monte Cristo, to be finally lost in the swamps of Flat Creek.

In this enormous space and solitude another dimension of life can be felt and grasped by the ready mind—uncramped and unfettered but threatened now by that slim, rusty slash to be seen in the saddle below. A cabin on a distant lake, a cache here and a tent site there; waterhole, creek, and berry patch: of such native material can a life be made, and the land met on its own terms. A good life on earth, as such things go, keeping at its center an uncommon dignity.

I suppose that with the passing of such individuals a way of life has been dying that we will not see again on earth. The wilderness has been labeled and set aside; the vastness has been partitioned, the great American solitude broken for what may be the last time. Now it is refuge, park, or preserve; or it is one more entry on the list of resources, in the dubious terminology of our times. Fortunately, the timber on these hills is of little commercial value.[1] It occurs to me that it was considerate of God to make the trees so small; at least our agencies won't come here to make a million in timber sales.

———

My affection for this landscape has been among the oldest and deepest things in my life. This is where I was born, improbably, at the age of twenty-three. Then I had come north to homestead, to till the soil, to pioneer. The years have tempered my zeal for pioneering, though I concede that for Alaska to support a people it must be able to provide them with food, that basic substance of soil and water. And I believe it can do so, with wise practice and understanding, though I suspect with diminished expectations in the years ahead.

Though I am not a farmer, I have cleared my plot of ground, broken up the sod and planted my seeds, as much as I needed to. And through necessity, nourishing my crops with hoe and watering can, I have learned my lessons with hard labor. I respect the honest man of the soil, even if he works it with a machine. I have seen the cleared fields bordering the road to Chena Hot Springs outside Fairbanks, the crops ripening under the sun. The barley fields in the Clearwater district near Delta have opened the forest and let in the wind and light, and that sight is beautiful to me in its own right.

It is the newness of these things here that confounds us, the swiftness of impending changes, in which we sense more keenly the immediate losses. The intact and roadless wilderness we have known is disappearing, and we will not see it again. Here too in Alaska and the North we will learn to live with diminished horizons. For now, we must care for what we have. There is no way to make it larger.

———

I look west into the light that comes across the hills, toward Banner Dome, four miles away. I have been told that seventy or more years ago, on a summer holiday, the townspeople of Richardson would climb that dome by foot and horseback to picnic in the open meadows. I have imagined them up there, in full skirts, bonnets, and broad hats, picking blueberries, and excited by the view.

Looking out on these hills today one would hardly guess the activity of those years—the numbers of men and animals that the country contained, the creaking sleds and groaning wagons. The trails we walk today were the roads they made through the wilderness, surveyed by eyesight and cleared with an axe. And God knows they did their share of harm in the country, with deliberately set fires, clearcut slopes, and scoured hillsides. They were years of hardship and deprivation, of crudity and innocence, of a thoughtless taking of the land's abundance, but years of easy and exuberant companionship, also, if one can believe the written accounts. And now the hills sleep in the sun, and there is not a sound to be heard above the wind.

It is just past 9 P.M. The sun is a long time going, settling in the long northwesterly sweep. While the light is still strong, we decide to take a short walk over the big meadow rising behind us. We want to see something of the country hidden from us by the south ridge of the dome.

The wind sweeps in a big rush across the tundra, pushing us on. There are flowers underfoot: avens, harebells, a small Jacob's ladder, a variety of cinquefoil, others I do not know.

In about twenty minutes of walking we are clear of the south shoulder, and stop to look. A vast panorama of mountain, plain, and riverbed stretches before us: the Tanana with its many islands, its sandbars and gleaming gray channels, and beyond it the irregular, snow-streaked profile of the Alaska Range.

Looking below us, we see the slim, geometrical cut of the Richardson Highway running east toward Delta Junction. And then, in random succession, the Granite Mountains, Donnelly Dome, Isabel Pass, and the sandy, braided wash of the Dry Delta descending from one of the glaciers on Mount Hays. In the middle distance, the big, blue oval of Quartz Lake and the level, dark green expanse of Shaw Creek Flats with its countless ponds.

The night shadows are spreading slowly over all that country, deepening the color of lake, river, and sky, though a misty, orange glow still bathes the farther hills toward the Goodpaster River. The view commands a stillness in us. Any steadfast look beyond the arrangements of men can return us to an earlier world, and it is as if no work had ever been done. And for a moment I imagine I understand how it must have been for the gods looking down from Heaven or high Olympus, gazing with detachment and concern on a far place of road and township, that strange, disruptive haunt of the tribes of men.

It is all very grand, and if we had time and shelter at hand it might be well to watch through the few short hours of the night until the sun rises. But the edge of the meadow where we are standing offers no shelter from the wind that freshens from time to time, hitting us with a stronger gust. We are tired from the day's walk, and having seen part of what we came for, we return to our camp.

————

There are times when high in the uncomplicated air of these domes and hills I feel like voicing a loud complaint against the persistent follies of my kind. Damn that fellow, Prometheus, anyway! Why couldn't he have left well enough alone? Why not have left humankind to perish in darkness and cold, as almighty Zeus intended? But no, he had to bring them fire, and stolen fire at that. And look what we've done with it, and all the gifts that followed. We have the arts, the crafts, the sciences, the engineering triumphs, the high achievements—yes. But also the wars and the waste, the compounded vanity and misery, the endless criminality, the destruction of landforms, the decimation of species. And who knows? The gods may win in the end, and a few debased survivors of our kind will perish at last, if not on earth, then in some miserable black hole in frigid space, in the darkness and cold held off for so long.

And would the wilderness have missed us? Not at all, I think. These hills would still have their summers, their winter snows, their profusion of fruit and flowers, and the moose, the caribou, and the bright birds of passage would be part of a wonderful plenty.

And then I think that without that divine arrogance and interference we would not be here. I would not be here. And I guess that settles it. For I would not want to have missed being here in this wind and evening light, nor to have missed my days in this country.

————

It is past ten o'clock. And now the sun dips below the farthest domes we can see. The light is soft and golden, suffused over the hills in a prolonged twilight to be found only in these high latitudes.

We unroll our sleeping bags and arrange our bedding on the uneven, spongy tent floor. It is good, finally, to lie down. Light from the sun below the hills in the northwest comes into the tent. The wind billows the blue translucent walls, and the tent eaves occasionally tug and strain.

I lie awake for a short while. Images, voices from past years, spring into my mind, released from the day's attention:

July 10. Very hot today. 80 in the shade. I done a wash. Picked blueberries. Swallows make a lot of noise. So hot I could not sleep.

Here in these hills I have known a kind of stillness and agreeableness in myself. Is it entirely imaginary that the ground underfoot somehow transmits a character and energy to the person who walks upon it? And now I remember a condition for happiness as defined once by Goethe: We are happy when for everything inside us there is an equivalent something out there. And once, it may be, we lived in such a world, when any stone, tree, or shadow might harbor a speaking spirit—the world alive and sentient, responding to the imagination in shapes of terror and joy. Not this half-dead thing of project and statistic, whose quantified presence confounds us with a spiritual absence. If there is evil in the world, let it take visible form in the shape of a goat, faun, or centaur, or a dragon breathing fire, as terrible as you wish. Not this formless, invisible menace that haunts the crowd of modern people like the atmosphere of a plague, ready to blossom into violent sores.

The wilderness is out there, quiet under the brief, rose-gray twilight before the sun rises again. But the wilderness is in myself, also, like a durable shadow. I prowl my region of flesh, my forest of blood, muttering and sniffing, turning many times in search of my own best place.

The tent walls flap. The air blown into the doorway is fresh and cool. I drift into sleep.

———

By late afternoon of the following day we are back in Fairbanks, threading the holiday traffic. We stop at a pizza parlor on College Road, near the university. We sit quietly in the semi-gloom, waiting to be served, with music coming too loudly from a speaker in the room. I am aware of traffic on the road outside, of the voices of others in the room, and of a vague but persistent dislocation. I am still part way back in the hills, walking the trail, standing on the dome in the west wind, looking out.

At times it is all but impossible to see our towns and cities, our houses, our cars and roadways, as anything but an imposition, a cruelty done to the earth, and for which we will be punished. And yet this too is our world, and we are all its half-willing conscripts. I recall now something that Robert Marshall wrote in *Alaska Wilderness* after one of his marvelous hikes in the Brooks Range early in the 1930s:

Now we were back among people in Wiseman. In a day I should be back in Fairbanks, in two more in Juneau, in a week in Seattle, and the great, thumping modern world. I should be living once more among the accumulated accomplishments of man. This world with its present population needs those accomplishments. It cannot live in wilderness, except incidentally and sporadically.

Well said and, probably, all too true. But would he have written those words in quite the same tone had he lived through the following decades and witnessed the corrosive effect of another world war and the relentless degradation of the planet? There is an innocence in those words, understandable and forgivable, for us who are the inheritors of those "accumulated accomplishments" that threaten us now as they have never before.

Meanwhile we drink our beer, brewed no doubt with mountain water, or with water pumped from some slowly emptying aquifer far under the paved foundations of our thumping modern world.

I think of the sun-lighted hills back there, of Campbell, and of a summer long past: one man alone in the country, taking his rare ease from labor in a camp by a shallow lake:

June 22. Took a rest. Listened to the birds. A lot of noise around here. I have a mountain bluebird out the door. Old cow moose in the lake. Everybody worked all night, except me and the dogs.

Note

1. This statement appears no longer to be true. The State of Alaska is now looking for ways to expand hardwood cutting in interior forests, offering to corporations like Fibre-form permission to cut thousands of board feet of birch and aspen. See Senate Bill 310.

Reference

Marshall, Robert. *Alaska Wilderness*. Berkeley: University of California Press, 1970.

FRAMEWORKS

Except possibly his soul, man prizes his mind above all else. His mind is a product of his ecology as much as the rest of him. Every human mind is a product of its ecology—the same ecology. Nothing that evolves persists unless sustained by those same creative forces. Like a ball at the top of a fountain, the human head pivots on its animal backbone, the mind a turning knot of thought and dream on the end of a liquid spear of living animals.

—*Paul Shepard*, Thinking Animals

A distinguishing trait of our species is the desire to construct frameworks. These are the clotheslines upon which we arrange the apparel of our understandings of place and our place in the order of things. On the one hand, frameworks are constructs that attempt to incorporate and mimic strands of the natural world; on the other, they are particularly human forms of architecture precipitated from the streambed of logos.

Frameworks explore how we attach meaning to things—how culture informs those meanings and steeps them in metaphor. Analytic discourse is a means of resolving a whole into parts, a way of looking back, and out, and over the broad terrain and discovering the relations which hold things together. The contributions to this section explore many of the diverse ways humans conceive of wildness and wilderness and the consequences of those varied perceptions.

It is always useful to keep Wallace Stegner's admonition in mind (an advisory which bears a close resemblance to one issued by Aldo Leopold when he suggested the need to keep all the parts): "The dreams of man," Stegner said, "will not accept what nature hands us. We have to tinker with it, trying to give it purpose, direction and meaning—or, if we are of another turn of mind, trying to demonstrate it has *no purpose, direction, or meaning. Either way, we can't let it alone."*

Each time we attempt to reconstruct nature we end up creating "products" of our own invention and have a difficult time seeing our way out of the epistemological boxes we have constructed. Living in a human-invented environment, reacting solely to our own creations, leaves us living entirely within our own minds. Restraint and humility must always temper our presumed knowledge of natural systems. In an essay called "Sanctity and Adaptation," Roy Rappaport has written: "It is perhaps the case that knowledge will never be able to replace respect in man's dealings with ecological systems, for . . . the ecological systems in which man participates are likely to be so complex that he may never have sufficient comprehension of their content and structure to permit him to predict the outcome of many of his own acts."

Scattered Notes on the Relation Between Language and the Land

DAVID ABRAM

For some time I have been working on a book exploring the ecology of perception and language, analyzing the manner in which these two dimensions modulate the relation between human culture and the living land. One of several intertwined themes in this project concerns the influence of writing—and, in particular, the alphabet—upon the experience of language and, consequently, upon our sensory experience of the earth around us. How has writing altered our awareness of time, of space, of earthly place? Here, in keeping with the focus of an anthology on culture and wildness, I have gathered from that project some scattered notes concerning the forgotten intimacy between language and the animate earth.

—————

A living language is continually made and remade, woven out of the silence by those who speak. And this silence is that of our wordless participations, of our perceptual immersion in the depths of an animate, expressive, more-than-human world.

—————

Expressive, living speech is a gesture—a vocal gesticulation wherein the meanings are inseparable from the sound, the shape, and the rhythm of the words. Communicative meaning is always, in its depths, affective; it remains rooted in the sensorial and bodily dimension of experience, born of the body's native capacity to resonate with other bodies and with the landscape as a whole. Linguistic meaning is not some ideal and bodiless essence that we arbitrarily assign to a physical sound or word and then toss out into the "external" world. Rather, meaning sprouts in the very depths of the sensory world—in the heat of meeting, encounter, participation. We do not, as children, first enter into language by consciously studying the formalities of syntax and grammar or by memorizing the dictionary definitions of words, but rather by actively making sounds—by crying in pain and laughing in joy, by squealing and babbling and playfully mimicking the surrounding soundscape, gradually entering through such mimicry into the specific melodies of the local language, our resonant bodies slowly coming to echo the inflections and accents common to our locale and community.

We thus learn our native language not mentally but bodily. We appropriate new words and phrases first through their affective tonality and texture, through the way they feel in the mouth or roll off the tongue, and it is this direct, felt significance—the *taste* of a word or phrase, the way it affects or modulates the body—that provides the fertile, polyvalent source for all the more refined and rarefied meanings which that term may come to have for us. Linguistic meaning remains rooted in the sensory life of the body—it cannot be completely cut off from the soil of direct, sensuous experience without withering and dying.

Yet to affirm that linguistic meaning is primarily gestural, and poetic, and that conventional and denotative meanings are inherently secondary, and derivative, is to renounce the claim that language is an exclusively human property. If language is always, in its depths, essentially carnal, and expressive, then it can never be definitively separated from the expressiveness of birdsong or the evocative howl of a wolf late at night. The chorus of frogs gurgling in unison at the edge of a pond, the snarl of a wildcat as it springs upon its prey, or the distant honking of Canadian geese vee-ing south for the winter—all reverberate with affective, gestural significance, the same significance that vibrates through our own conversations and soliloquies, moving us at times to tears, or to anger, or to intellectual insights we could never have anticipated. Language as a bodily phenomenon accrues to all expressive bodies, not just the human. Our own speaking, then, does not set us apart from the animate landscape but—whether or not we are aware of it—inscribes us more fully in its chattering, whispering, soundful depths.

If, for instance, one comes upon two human friends meeting for the first time in many months and chances to hear their initial words of surprise, greeting, and pleasure, one may readily notice, if one pays close attention, a tonal, melodic layer of communication beneath the explicit denotative meaning of the words—a rippling rise and fall of the voices in a sort of musical duet, rather like two birds singing to each other. Each voice, each side of the duet, mimes a bit of the other's melody while adding its own inflection and style and then is echoed by the other in turn—the two singing bodies thus tuning and attuning to each other, rediscovering a common register, *remembering* each other. It requires only a slight shift of focus to realize that this melodic singing is carrying the bulk of communication in this encounter. The explicit meanings of the actual words ride on the surface of this depth like waves on the surface of the sea.

It is by a complementary shift of attention that one may suddenly come to hear the familiar song of a blackbird or a thrush in a new manner—not just as a pleasant melody repeated mechanically, as on a tape player in the background, but as active, meaningful speech. Subtle variations in the tone and rhythm of those whistling phrases suddenly seem laden with expressive intention, and the two birds singing to each other across the field appear for the first time as attentive, conscious beings, earnestly engaged in the same world that we ourselves engage, yet from an astonishingly different angle and perspective.

———

As there are fewer and fewer songbirds in the air, due to civilization's destruction of their habitat and the tropical forests where many of them winter, it is likely that our own speaking is losing some of its melodic resonance and power. For if our speech no longer echoes and responds to the voices of the warbler and wren, it can no longer be informed by their cadences. As the splashing speech of the rivers is silenced by more dams, as we drive more and more of the land's wild voices into the oblivion of extinction, our own words become increasingly impoverished and weightless, progressively emptied of expressive meaning.

ON ORAL CULTURE

To oral peoples, nature itself is articulate: it *speaks*. The human voice in an oral culture is always to some extent participant with the voices of wolves, wind, and waves—participant, that is, with the encompassing discourse of an animate earth. There is no element of the landscape that is definitively void of expressive resonance and power: any movement may be a gesture, any sound may be a voice, a meaningful utterance.

———

In the oral, animistic world of pre-Christian and peasant Europe, all things—animals, forests, rivers, and caves—had the power of expressive speech, and the primary medium of this collective discourse was the invisible air. In the absence of writing, human utterance, whether embodied in songs, stories, or spontaneous sounds, was inseparable from the exhaled breath. (Try speaking a phrase without exhaling.) Spoken words, it was assumed, are formed breath and take their communicative power from the breath and the air. Hence the enveloping atmosphere was an implicit intermediary in all communication—a zone of subtle influences crossing, mingling, and metamorphosing. This invisible dimension of whiffs and scents, of vegetative emanations and animal exhalations, was also the deep repository of ancestral voices and intentions, the home of stories yet to be spoken, of ghosts and spirited intelligences—a kind of collective field of meaning from whence individual awareness continually emerged, and into which it continually receded, with every inbreath and outbreath.

We might say that the air, as the invisible wellspring of the present, yielded an awareness of transformation and transcendence very different from that *total* transcendence expounded by the church. The experiential interplay between the visible and the invisible—this duality entirely proper to the sensuous lifeworld—was far more real, for oral peoples, than an abstract dualism between sensuous reality as a whole and some other, utterly nonsensuous heaven.

―――――

Whenever we seek to engage the discourse of oral cultures, we must free ourselves from our habitual impulse to visualize any language as a static structure that can be diagrammed or a set of rules that can be listed. Without a formal writing system, the language of an oral culture simply cannot be objectified as a distinct entity by those who speak it. And this lack of objectification influences not only the way in which oral cultures experience the field of discursive meanings but also the very character and structure of that field. In the absence of any written analogue to speech, the sensible, natural landscape remains the primary visual counterpart of spoken utterance, the visible accompaniment of all spoken meaning. The land, in other words, is the sensible site or matrix wherein meaning occurs and proliferates. We are situated in the field of discourse as we are embedded in the natural landscape—the two matrices are not separable—and we can no more stabilize the language and render its meanings determinate than we can freeze all motion and metamorphosis within the land.

If we listen, first, to the sounds of an oral language, to the rhythms, tones, and inflections that play through the speech of an oral culture, we are likely to find that these elements are attuned, in multiple and subtle ways, to the contour and scale of the local landscape, to the depth of its valleys or the open stretch of its distances, to the sensuous rhythms of the local topology. But the human speaking is necessarily tuned, as well, to the various nonhuman calls and cries that animate the local terrain. Such attunement is simply imperative for any culture still dependent upon foraging and hunting for its subsistence. Minute alterations in the weather, changes in the migratory patterns of prey, a subtle shift in the focus of a predator—sensitivity to such subtleties is a necessary element of all oral subsistence cultures, and this sensitivity is inevitably reflected not just in the content but in the very shapes and patterns of human discourse.

Hunting for food, within an indigenous community, entails abilities and sensitivities very different from those associated with hunting in technological civilization. Without guns and gunpowder, a native hunter must come much *closer* to his prey if he is to take its life—not just physically but emotionally, empathically entering into proximity with the other animal's ways of sensing and experiencing. The native hunter, in effect, must apprentice himself to the animals he would kill. Through long and careful observation—enhanced, at times, by ritual identification and mimesis—the hunter gradually develops an instinctive knowledge of the habits of his prey, of its fears and its pleasures, its preferred foods and favored haunts. Nothing is more integral to this practice than learning the communicative signs, gestures, and cries of the local animals. Knowledge of the sounds by which a monkey indicates to the others in its band that it has located a good source of food, or the cries by which a particular bird signals distress, or by which another attracts a mate, enables the hunter to anticipate both the large-scale and small-scale movements of these animals. A familiarity with animal calls provides the hunter, as well, with an expanded set of senses—an awareness of events happening beyond his field of vision, hidden by the forest leaves or obscured by the dark of night. Finally, the skilled human hunter can also generate and mimic such sounds himself, and it is this gift that enables him to enter most directly into the society of other animals. It would be very strange if such sounds and signs, so crucial to our indigenous ancestors, had not informed the genesis and early development of all our human languages.

———

Without a versatile writing system there is simply no way to preserve, in any external medium, the accumulated knowledge regarding particular plants (including where to find them, which parts are edible or poisonous, how they are best prepared, what ailment they may cure, or exacerbate), regarding specific animals (where they hang out, what they eat, how best to track or hunt them), or even regarding the land itself (how to orient oneself in the surrounding terrain, what landforms to avoid, where to find water or fuel). Such practical knowledge must be preserved, then, in spoken formulations which can be easily remembered, modified when new facts are learned, and retold from generation to generation. Yet not all verbal formulations are amenable to simple recall—most verbal forms that we are conversant with today are dependent upon a context of writing. To us, for instance, a simple mental list of the known characteristics of a plant or an animal would seem the most obvious formulation. Yet such a list has no value in an oral culture—for without a visualized counterpart which can be brought to mind and scanned by the mind's eye, such lists cannot be readily recalled and repeated.

Within a nonwriting culture, knowledge of the diverse properties or powers of other animals, plants, and the land itself can be remembered and preserved only by being woven into *stories*—into vital tales wherein the animal's characteristics are made evident through a narrated series of events and interactions. Stories, like rhymed poems or songs, readily incorporate themselves into our felt experience; the shifts of action echo and resonate our own encounters. Hearing or telling a story we vicariously *live* it, and the encounters and travails of its characters embed themselves into our own flesh.

The sensuous, breathing body is a dynamic, ever-unfolding form—more a process than a fixed or unchanging object. As such, it cannot readily appropriate inert "facts" or "data" (static nuggets of "information" abstracted from the situations in which they arise). Yet the living body can assimilate stories, approaching each episode as a variation or distant extension of its own dynamism. And the more lively the story—the more vital the encounters within it—the more readily it will be incorporated. Oral memorization calls for dynamic (often violent) characters and encounters. If the story carries knowledge about a particular plant or natural element, that entity will often be cast, like all the other characters, in a fully animate form, capable of personlike adventures and experiences, susceptible to the kinds of setbacks or difficulties we know from our own lives. In this manner the *character* or personality of a medicinal plant will be easily remembered, its poisonous attributes will be readily avoided, and the precise steps in its preparation will be evident from the sequence of events in the very legend one chants while preparing it. One has only to recite the appropriate story, from the distant time, about a particular plant, animal, or element in order to recall the accumulated cultural knowledge regarding that entity and its relation to the human community.

ON STORY

An important reason for the profound association between stories and the more-than-human terrain resides in the encompassing wholeness of a story in relation to the characters that act and move within it. A story envelops its protagonists much as we are ourselves enveloped by the landscape. In other words, we are situated in the land in much the same way that characters are situated in a story. For members of a deeply oral culture, this relation may be experienced as something more than mere analogy: along with the other animals, the stones, the trees, and the clouds, we are ourselves characters within an imaginative story that is visibly unfolding all around us, participants within the vast imagination, or dreaming, of the earth.

To hear a story told and retold in one's childhood, and to recount that tale when one has earned the right to do so (now inflected by the patterns of one's own experience and the rhythms of one's own voice), is to actively preserve the coherence of one's culture. The practical knowledge, the moral patterns and social taboos, and indeed the very language or manner of speech of a nonwriting culture maintain themselves primarily through narrative chants, myths, legends, and trickster tales—that is, through the telling of stories.

It is clear, moreover, that the stories of an oral culture are deeply bound to the earthly terrain inhabited by that culture. The stories, that is, are profoundly and indissolubly place-specific. The Distant Time stories of the Koyukon, the Agodzaahi tales of the Western Apache, and the Dreaming stories of the Pintupi and Pitjanjarra present three very different ways whereby tribal stories weave the people who tell them into their particular ecologies—or, still more precisely, three ways in which earthly locales may speak through the human persons that inhabit them. The telling of stories, like singing and praying, would seem to be an almost ceremonial act, an ancient and necessary mode of speech that tends the earthly rootedness of human language. For narrated events always *take place*, always happen *somewhere*. And for an oral culture, that locus is never merely incidental to the occurrences. The events belong, as it were, to the place. And to tell the story of the events is to let the place itself speak through the telling.

The stories of such cultures give evidence, then, of the unique power of particular bioregions—the unique ways in which different ecologies call upon the human community. But these stories also give evidence of the character of specific sites, or places, within that larger region. The specificity, the singular magic, of a place is evident from what happens there, from what befalls oneself or others when in its vicinity. To tell of such events is implicitly to tell of the power of that site and indeed to participate in its expressive potency. In these cultures, to tell certain stories without saying precisely where the events occurred (or, if one is recounting a vision or dream, to neglect to say where one was when granted the vision) may alone render the telling powerless or ineffective.

The songs proper to a site will share a common style—a rhythm that matches the beat or the pulse of the place, attuned to the way things happen there, to the sharpness of the shadows or the rippling speech of the water bubbling up from the ground. In Ireland, a country person might journey to one distant spring in order to cure her insomnia, to another spring for strengthening her ailing eyesight, and to yet another to receive insight and protection from thieves. For each spring has its own powers, its own blessings, its own curses. Different gods, and different demons, dwell in different places. Each place has its own dynamism, its own pulse and patterns of movement, and these patterns engage the senses and relate them in particular ways, instilling particular moods and modes of awareness—so that unlettered, oral people will rightly say that each place has its own mind, its own personality, its own intelligence.

ON WRITING

Writing, like human language, is engendered not just within the human community but between the human community and the animate landscape, born of the interplay between the human and the more-than-human world. The earthly terrain is shot through with suggestive scrawls and traces—from the sinuous calligraphy of rivers winding across the land, inscribing arroyos and canyons into the parched earth of the desert, to the black slash burned into the trunk of an old elm by lightning. The swooping flight of birds is a kind of cursive script written on the wind; it is this script that was studied by the ancient augurs, who could read therein the course of the future. Leaf-miner insects make strange hieroglyphic tabloids of the leaves they consume. Wolves urinate on specific stumps and stones to mark off their territory. And today you read these printed words as tribal hunters once read the tracks of deer, moose, and bear printed in the soil of the forest floor. Archaeological evidence suggests that for more than a million years the subsistence of humankind has depended upon the acuity of such hunters, upon their ability to read the traces—a bit of scat here, a broken twig there—of these animal Others. These letters across the page, tracking across the white surface, are hardly different from the footprints of prey left in the snow. We read these traces with organs honed over millennia by our tribal ancestors, moving instinctively from one track to the next, picking up the trail afresh whenever it leaves off, hunting the meaning, that is, the meeting with the Other.

———

Perhaps the most succinct evidence for the potent magic of written letters is to be found in the ambiguous meaning of our common English word *spell*. As the Roman alphabet spread through oral Europe, the Old English word *spell*, which had meant simply to recite a story, took on a new double-meaning: on the one hand it now meant to arrange, in the proper order, the written letters that constitute the name of a thing or person; on the other it signified a magic formula or charm. Yet these two meanings were not nearly so distinct as they have come to seem to us today. For to arrange the letters which make up the name of a thing, in the correct order, was precisely to cast a magic spell—to establish a new kind of influence over that entity, to summon it forth. To spell—to correctly arrange the letters to form a name or a phrase—seemed thus at the same time to cast a spell: to exert a new and lasting power over the things spelled. Yet to learn to spell was also, more profoundly, to step under the influence of the written letters oneself: to cast a spell upon one's own senses. It was to exchange the wild and multiplicitous magic of an intelligent natural world for the more concentrated and refined magic of the written word.

———

As the technology of writing spreads through a previously oral culture, the power and personality of certain places begin to fade. For the stories which express and embody that power are gradually recorded in writing. Writing down oral stories renders them separable, for the first time, from the actual places where the events occurred. Stories, which arose in the interaction between people and the land they inhabit, can now be carried elsewhere; they can be read in distant cities or even on alien continents. The stories, soon, come to seem independent of any particular locale.

Previously, the power of spoken tales was necessarily rooted in the power of places where their events unfolded. While the telling of certain stories might be provoked by specific social situations, their instructive value and moral efficacy was often dependent upon one's sensible contact with the sites where those stories happened. Other stories might be provoked by a direct encounter with a species of bird or animal whose exploits figure prominently in the tales, or with a particular plant just beginning to flower, or by local weather patterns and seasonal changes. In such cases, contact with the regional landscape (and the multiple sites or places within that landscape) was the primary mnemonic trigger of the oral stories and thus was integral to the preservation of the stories and indeed the culture itself.

Once the stories are written down, however, the visible text becomes the primary mnemonic activator of the spoken stories—the inked traces left by the pen as it traverses the page replacing the earthly traces left by the animals and, as well, by one's ancestors in their interactions with the land. The places themselves are no longer necessary to the remembrance of the stories. Often they come to seem wholly incidental to the tales—the arbitrary backdrops for human events that might just as easily have happened elsewhere. The transhuman ecological determinants of the originally oral stories are no longer emphasized; often they are written out of the tales entirely. In this manner the stories and myths, as they lose their oral character, forfeit as well their intimate links to the more-than-human earth. And the land itself, stripped of the particularizing stories that once sprouted from every cave and streambed and cluster of trees, begins to lose its multiplicitous power. The human senses, intercepted by the written word, are no longer gripped and fascinated by the expressive shapes and sounds of particular places. The spirits fall silent. Gradually, the felt primacy of place is forgotten, superseded by a new, abstract notion of "space" as a homogeneous and placeless void.

———

It should be easy, now, to understand the destitution of indigenous oral people who have been forcibly displaced from their traditional lands. The local earth is, for them, the very matrix of discursive meaning. To force them from their native ecology (for whatever political or economic purpose) is to render them speechless—or render their language meaningless—and dislodge them from the very ground of coherence. The massive "relocation" or "transmigration" projects under way in numerous parts of the world today in the name of "progress"—for example, the forced "relocation" of oral peoples in Indonesia and Malaysia to make way for the commercial clearcutting of their forests—must be understood, in this light, as instances of cultural genocide.

For the great majority of indigenous, traditionally oral peoples, the coherence of human language is inseparable from the coherence of the surrounding ecology and the expressive vitality of the more-than-human terrain. It is the animate earth that speaks; human speech is but a part of that vaster discourse.

The Idea of Wilderness as a Deep Ecological Ethic

MAX OELSCHLAEGER

E. O. Wilson argues that the earth verges on a sixth mass extinction of life.[1] There is more than a little reason, then, to reconsider the idea of wilderness apropos the issue of protecting biodiversity. Clearly and unequivocally, the survival of the vast and rich array of life on earth depends upon the conservation of core habitat that remains unhumanized, that is, wild.[2] Wilson himself holds out some chance that bioeconomic analysis might allow some portion of, for example, the Amazon rainforest to be developed consistent with the end of protecting biodiversity; he also insists that large portions of undeveloped lands must stay wild in perpetuity.

Regrettably Wilson is in a small minority among conservationists, since they argue, under the banner of sustainable development, that humankind can manage planet Earth.[3] Of course, not everyone means the same thing by sustainable development; with careful definition it might be a workable concept. Yet as used within mainstream environmentalism, sustainable development is an oxymoron, scientifically dubious,[4] and philosophically empty.[5] More than anything else, sustainable development intends to substitute a prosthetic environment for wild nature and ecotechnologies for the biophysical processes that sustain life.[6]

The creation of a prosthetic environment is an illusory goal that rests upon a "surface ecological ethic" or "shallow ecology" that ignores the stubborn reality that humankind is inextricably embedded within biophysical process. "Only in the last moment of human history," Wilson argues, "has the delusion arisen that people can flourish apart from the rest of the living world." But through biophilia, the love of life, and more generally through the idea of wilderness, Wilson claims, we might develop a deep conservation ethic that naturalizes history and reconnects us with earth. "Wilderness," he continues, "settles peace on the soul because it needs no help; it is beyond human contrivance."[7] It reminds us of our place in a larger scheme of things.

In this essay I briefly develop six aspects of the idea of wilderness in relation to the theme of the protection of biodiversity. These are: practical, psychological, political, ethical, ecological, and cosmological. Together these various aspects of the idea of wilderness indicate a path toward a deep ecological ethic that is also a practice—a way of life. Following the lead of Arne Naess, we might call these considerations "Ecosophy W," although a complete statement of such a philosophy is not offered here.[8]

PRACTICAL ASPECTS

Beyond almost any reasonable doubt, conservation and management of wilderness is the most feasible strategy for the protection of biodiversity—in short, the surest means by which a sixth mass extinction of life might be avoided.[9] While a number of reasons might be adduced in defense of my conjecture, the basic reasons are two: knowledge and capital.

Consider that humankind lacks the requisite knowledge to replace nature's economy with a prosthetic environment.[10] In the first place, the advance of knowledge cannot be predicted; even if technological fixes were desirable, a dubious proposition in its own right, there are no guarantees that the theoretical knowledge necessary to engineer them would be forthcoming. Further, it is the height of arrogance to believe that humankind can substitute its plans and so-called ecotechnologies for the biophysical processes that evolved over billions of years of geological and biological time. Indeed, given the seemingly intractable increase in the measures of ecocrisis (extinctions of species, human population increases, destruction of rainforests, and more), the preponderance of evidence implies that we presently do not know how to manage the planet. And even if we did know how, we would collectively be unable to do so.[11]

That mainstream environmentalism would climb on the managing planet Earth bandwagon is comprehensible, since the last several centuries have been predicated on relentless economic development, technological advance, and the idea of progress.[12] So framed, the development of atomic power and the exploration of space (to take examples of technologically altered relations to the microcosmic and macrocosmic universe) predispose reform environmentalists to think that human beings have the savoir faire to manage anything and everything, including global ecology. Yet there is little evidence that we can do so. Indeed, given the rapidly deteriorating condition of the biosphere, it appears that we are better at disrupting than at conserving the integrity and stability of life.

There are several reasons to carefully consider alternatives before going any further down the path of expert management for so-called sustainable development.[13] For one, the path of scientific discovery and technological development (or engineering) is never as predictable as the experts claim; it is always pockmarked with delays, failures, and even catastrophes. And technological fixes often fail to solve the problems they address. Worse, they sometimes create new problems. Finally, and perhaps most important, the kind of world that the technological imperative leads toward is potentially antithetical to our most basic values.

Even assuming that a totally managed environment is feasible, which incorrectly assumes that the requisite knowledge to create a prosthetic environment is in hand, the question of capital complicates the issue. Consider stratospheric ozone depletion: the entire fleet of the world's jumbo jets, flying twenty-four hours a day for seven days a week, week in and week out, could not carry enough ozone to fill in the "holes" rending the ozone layer. Furthermore, the economics of the technological path ensures continuing marginalization of the Third World. The Japanese plan to become the world leader in producing and marketing the ecotechnologies of tomorrow; such technologies will not come cheaply. Consumers of these technologies will pay not only the costs of development and marketing; they will fund profits as well. The irony in the quest for the technological fix, as Vandana Shiva (and other Third World critics) makes abundantly clear, is that reform environmentalism assumes that all people in all places want to become consumers living like Europeans and Americans.[14]

To sum up my first point: the idea of wilderness stands as a theoretical alternative to the managing planet Earth philosophy. It is practical to conserve wild nature—not only because we continue to rely upon the tested and confirmed wisdom of nature's own biophysical processes, but because we do not have to rely upon the production of human knowledge (such as biotechnology) and because the basic capital is nature's own.

PSYCHOLOGICAL ASPECTS

More than one critic of mainstream environmentalism has argued that if we are to conserve life on earth we must somehow develop an alternative definition of human beingness.[15] Clearly this is a complicated subject, one that I can only begin to discuss here. Yet fundamental questions about the human psyche are sharpened by an encounter with the idea of wilderness in the context of biodiversity.[16].

One of the problems with the culturally dominant (Eurocentric) conception of ourselves is that we are defined as beings who exist outside of nature. Descartes' famous philosophical dictum, *cogito ergo sum*, suggests at least part of the idea; on his account, we are thinking things without bodies. What is lost is any notion that we are deeply and abidingly physical beings, that we are of and about earth. In Descartes' conception we are, through the power of science and technology, the masters of nature.[17]

But to grasp the idea that we are of and about earth brings home clearly that our second nature is a veneer, a cultural accretion that overlays a wild nature, a first nature. To come to grips with this biological being of humanity is to begin to sense a deeper affiliation with life on earth and to realize that our conventional notion of the self is an artifice, a contingency.[18] This created self, serving as a distinguishing criterion for Western Man, is a psychological foundation for the modern environmentalist: Man-who-would-manage-the-planet—grown powerful and important through the prestige of award and the status of appointment.

But who is it that speaks? Who is this Man who claims to see the future and to engineer sustainable development? He is the prototypical person of the modern age, the Man who stands astride history, committed to the continued economic and demographic expansion of the human species. And this Man is precisely the man brought into question by Aldo Leopold and Susan Griffin, Arne Naess and Vandana Shiva, and many others who have found reason to question the feasibility of managing the planet for sustainable development.

Caught up in our modern stories of history, technology, progress, economic growth, and development, our roots in earth have been concealed. Yet do we not, in our speaking—in the very noises we make, as we inhale and exhale, as we move our tongues upon our teeth, as we make modulated noises that are meaningful—confirm the still living bonds that tie us to earth? Think of the awesome birth of vociferation, as Maurice Merleau-Ponty so powerfully describes it, or the explosive origin of speech, to use Paul Ricoeur's phrase.[19] We are flesh, flesh of earth, speaking flesh of earth. Such a realization opens the possibility of redefining our place in the scheme of things.[20]

Ernest Becker argues that the denial of our physical being, of our roots in earth, manifests itself in the fear of death.[21] The relentless conversion of earth to the purposes of consumption is, according to Becker, one consequence—a reflection of unconscious and unresolved anxieties inherent in the human condition. He equates the denial of death with neurosis, a psychic malaise that distorts culture, twists the lives of individuals, and wreaks havoc upon the biosphere. In Becker's analysis there is no solution for ecocrisis, and perforce no possibility of a deep conservation ethic, until we rethink the human position in the natural scheme of things.

So contextualized, the denial of death is a psychopathology impelling humankind along a self-defeating path since, among other consequences, it sustains the prevailing cultural definition of ourselves as exclusively historical beings—that is, as somehow existing entirely above or outside biophysical process.[22] Wilderness experience, as John Muir and many others note, helps to counteract the socially conditioned fear of death. Stabilized into an idea of wilderness, such experience prompts humans to slip the noose of the death orthodoxy by reconceptualizing life and death as successive moments of a single cosmic scheme. "Life seems neither long nor short," Muir writes, "and we take no more heed to save time or make haste than do the trees and stars. This is true freedom, a good practical sort of immortality."[23]

POLITICAL ASPECTS

Political aspects of the idea of wilderness cover a gamut of issues—from the kind that conservatives like Ayn Rand and radicals like Edward Abbey address in their concern for wilderness areas as a haven for radicals who oppose the power of the modern state to questions of equity between First and Third World nations. My remarks here are limited to a brief point apropos the global politics of biodiversity.

Given the historical stage upon which we can stand, there can be little doubt that the world is governed by a small number of powerful nation-states whose interests are conjoined with those of economically powerful and politically influential multinational corporations. Acting conjointly, and often with the tacit cooperation of the United Nations (funded almost totally by the superpowers), this axis of power promotes the process of globalization and sustainable development.

Globalization reflects the status quo: more than anything, it is the attempt to extend the processes of industrialization and mass consumption across the entire world. Paul Kennedy argues, in *Preparing for the Twenty-First Century*, that nation-states themselves are losing power in the postmodern world to the multinational corporations that control the global marketplace, communication networks, international banking, and processes of technological research and design. Perhaps he is correct. In any case, globalization is a relentless project that refuses to reckon with ecological reality. Indeed, this reality is concealed from most of the world's people under the veneer of "sustainable development," a euphemism that hides the naked truth—namely, the fact, as E. O. Wilson puts it, that the human species is ecologically abnormal.[24]

What sustainable development means, stripped of its euphemistic veneer, is strong anthropocentrism. What continues are the processes of human population growth and the inevitable humanization of the earth to promote rising levels of consumption, far beyond the levels required to meet vital needs.[25] Globalization, in any guise, leads to a world that is totally humanized, where the land is put to so-called productive purposes (as if wildlands are unproductive, as if biosystem services do not figure in the economic calculus of value) and where indigenous peoples are displaced to the margins by the center of power—namely, the "banks, corporations, speculators, governments, development agencies, and foreign power groups" that set the agenda for sustainable development.[26] The irony is that native peoples are displaced from the land and thereby economically marginalized by the so-called process of economic development.[27] The insult is compounded by the reality that the developed world transfers its most ecologically devastating processes of industrial production to the Third World, so that the consumerist lifestyles of Americans and Europeans can be enjoyed in environments that remain relatively pristine (even if intensely humanized).[28]

Is globalization leading us to the kind of world in which we want to live? Or is it leading toward administrative despotism and cultural homogenization which assumes that the Western consumer is the normative model for humankind? Even assuming that technology could replace the natural cycles that govern the global climatological and hydrological cycles (a major assumption in itself), to implement such a technology would require a new kind of social system. Murray Bookchin terms it "ecological technocracy." Such a system, he says, will "require a highly disciplined system of social management that is radically incompatible with democracy and political participation by the people."[29]

But what, if any, are the alternatives to the politics of the present—the power structure that promotes globalization and the interests of multinational corporations? I offer the barest sketch of an answer here: humankind must seriously consider reinhabiting the natural world. Although the U.S. Wilderness Act of 1964 defines wilderness as areas devoid of humans, wild nature was not historically devoid of human beings. Indeed, as I have argued elsewhere, humankind literally inhabited wild nature (and lacked any conception of a civilization as an ideal to which they aspired) for tens of thousands of years. For such people "wild" nature was home (suggesting that the kind of culture in which we live, which defines itself in opposition to the wilderness, is an enormous contingency).[30] Even today there remain examples of indigenous people living in ecosystems that support biodiversity (and sometimes, as ethnobotanists tell us, living in ways that actually increase biodiversity). We have much to learn, bioregionalists argue, by studying the ways of indigenous people. Such lessons might inspire the reform of our own culture, as we once again aspire to live in place.[31]

Politically, then, the idea of wilderness leads us toward the idea of a wild culture—a bioregional form of existence that challenges international modernism and the process of globalization.[32]

ETHICAL ASPECTS

The managing planet Earth philosophy exemplifies what Naess calls "shallow ecology" and Wilson calls a "surface ethic": a conservation ethic predicated on the platitudes of human self-interest, continued population and economic growth, and a "healthy environment." What is required if we are to avoid mass extinctions of life, is a deep conservation ethic. Wilson argues that a deep ecological ethic can be predicated on the explicit knowledge and account of *biophilia*—love of the living earth with which we are irretrievably bound.[33] Regardless of any latent tendency toward biophilia, modern Westerners neither acknowledge nor celebrate their affiliation with life. Their connections to earth are concealed by history as the defining story of Man's Conquest of Nature. But this is a potentially fatal illusion, one that is revealed by reflection upon the idea of wilderness.[34]

As many commentators make clear, there is no one deep ecological ethic that all must accept. And yet, in some sense, the notion that "in Wildness is the Preservation of the World" informs any effort to conserve biodiversity. Aldo Leopold, the creater of what is perhaps the best-known ecological ethic, the land ethic, catches the ethical import of the idea of wilderness.[35] To grasp "the cultural value of wilderness boils down," Leopold argues,

to a question of intellectual humility. The shallow-minded modern who has lost his [or her] rootage in the land assumes that he [or she] has already discovered what is important; it is such who prate of empires, political or economic, that will last a thousand years. It is only the scholar who appreciates that all history consists of successive excursions from a single starting-point, to which man [and woman] returns again and again to organize yet another search for a durable scale of values. It is only the scholar who understands why the raw wilderness gives definition and meaning to the human enterprise.[36]

Wildness leads to wild talk and wild knowledge, forcing us to realize that there is a natural semiosis, a deep and abiding structure to the world, that is beyond human invention and artifice.[37] Through socialization we are caught up in our own cultural semiosis, a structure of meaning that deceives us. We accept its legitimacy without engaging in what Naess calls "deep questioning." But immediate experience in wild nature and later reflection on the idea of wilderness promotes that process of criticism.

The power of wildness lies in helping us renew ourselves—to imaginatively create and linguistically capture words that move us toward a genuinely sustainable society where humankind's existence does not entail the mass extinction of the others. Maurice Merleau-Ponty has caught this idea: "In a sense the whole of philosophy . . . consists in restoring a power to signify, a birth of meaning, or a wild meaning, an expression of experience by experience, which in particular clarifies the special domain of language. And in a sense . . . language is in everything, since it is the voice of no one, since it is the very voice of the things, the waves, and the forests."[38] Will Wright takes the idea of wilderness one step further, arguing that "language always constitutes our world, the world we know, the world we live in—not in the sense that the world is language, or that language controls the world, but in the sense that the structure of language must be an inherent part of any world we can know."[39]

ECOLOGICAL ASPECTS

In one sense, the ecological aspects of the idea of wilderness are on the other side of practicality. The idea of wilderness reminds us that humankind exists only because there is a natural order or deep structure to life on earth. Ecocrisis, which now verges on ecocatastrophe (a word that has already found its way into the newest dictionaries), confirms that the modern world has lost touch with nature's order. Mesmerized by our own cultural order, we jeopardize the biophysical underpinnings of civilization as we act out an ecological *danse macabre*.

Clearly, there are few if any wild ecosystems that remain unaffected by human activity. In truth, humankind has affected wilderness ecosystems through hunting and gathering since time immemorial. But the rise of agri-culture, followed by industrial civilization, has dramatically changed the scope of human influence. Jerry Franklin and Ed Bloedel suggest that the influences of prehistoric peoples were "elements in the long-term evolution of the presettlement ecosystems— present for hundreds, thousands, and even tens of thousands of years. The impacts of modern humankind are *not* of this type. . . . Uncontrolled, modern humankind's influences alter the historical direction and rate of ecosystem evolution."[40] Industrial civilization's impact on wild ecosystems can be categorized in terms of function (that is, the ability of an ecosystem to sustain ongoing biophysical process), structure (that is, the physical-spatial arrangement of the ecosystem), composition (that is, the makeup and distribution of the flora and fauna), and dynamics (that is, the evolutionary trajectory itself).[41] Collectively considered, the influence of modern humankind has been characterized by conservation biology as "the death of life."

The study of wild ecosystems has implications beyond conservation alone. Aldo Leopold was among the first to explore this aspect of the idea of wilderness. He believed that wild nature, insofar as it could be found, served as a yardstick by which the cumulative effects of human culture on the natural world might be measured. More recently a variety of ecologically minded thinkers such as Eugene P. Odum, the renowned systems ecologist, have argued that the earth is a bioregenerative system about which we know some things but not enough for "a clear understanding of how the whole thing works." The point of studying wild ecosystems is to learn as much as we can so that we might begin to ameliorate the deleterious consequences of industrial civilization and human population growth. The more we learn about wild nature the more we also grasp the reality that biosystem services—the web of life that includes the flora and fauna, the microorganisms and ecosystems, the hydrological and atmospheric cycles, and so on—are literally the foundation, the sine qua non, of human existence.[42] So construed, as Gary Snyder emphasizes, wild ecosystems communicate nature's norms for respectful living on the planet.[43] In our civilized arrogance we have forgotten our ecological manners, jeopardizing not only the survival of the vast majority of the others but our continued existence as well.

COSMOLOGICAL ASPECTS

Perhaps there is no better place to look for the cosmological implications of the idea of wilderness than in the work of Henry David Thoreau. Amid the swirling clouds on Ktaadn's ridge, just at the point where his idea of wilderness might have collapsed into New England transcendentalism, Thoreau firmly seized the idea of cosmic process. His words ring true for us, for amid the rocks on the ridge leading to Ktaadn's peak he discovered granitic truth: "The highest that we can attain to is not Knowledge, but Sympathy with Intelligence. . . . There are more things in heaven and earth than are dreamt of in our philosophy."[44] Cosmos, Thoreau intuited, is Heraclitean. And capital "K" knowledge, be it scientific, humanistic, philosophical, or theological, is a Parmenidean deception, since the human animal is bound with the ongoing wild process of evolution.[45]

Twentieth-century savants, like Ilya Prigogine, have scientifically corroborated Thoreau's intuitions. Prigogine argues that we are in the middle of a second scientific revolution that is transforming, as did the first scientific revolution, our self-conception, indeed, our notion of culture itself.[46] Becoming, as Thoreau felt, rather than Being, or activity rather than passivity, undergirds all that is, was, or ever will be—including our own existence. We now know that some fifteen billion years of evolutionary history undergirds the appearance of species *Homo sapiens* on a small planet orbiting an ordinary star in the Perseus arm of the Milky Way galaxy—itself accompanied by a number of other galaxies in the local cluster. We are latecomers to the evolutionary process, and we do not know all that we might care to know. What we do know is that we are bearing sentient witness to cosmic chaos, the wild process through which all that exists has come to be. We are bound with time—that is, with the historic character of cosmos.[47]

The Paleolithic mind perhaps intuited the truth of human finitude and the miracle and utter gratuity of existence. By all indications the modern mind is devoid of any similar sensibility. We have relentlessly pursued the vision of ourselves as Lord Man. In our project to manage planet Earth we behave with radical ingratitude—blind to the cosmic chaos that enabled and yet sustains human being. Apart from this cosmic wilderness we cannot account for even the possibility of our existence, let alone its reality. As Henry Nelson Wieman notes: "The thin layer of structure characterizing events knowable to the human mind by way of linguistic specification is very thin compared to that massive, infinitely complex structure of events, rich with quality, discriminated by the noncognitive feeling-reactions of associated organisms human and nonhuman."[48]

Immediate experience, and later reflection on that experience, reminds us, helps us to re-cognize, that the cosmos is a vast society of interacting organisms whose history stretches far back into the dimmest recesses of space-time, beyond any culturally constructed definition of self and society and beyond the grasp of the human intellect to articulate in scientific terms. Arne Naess argues: "The world as spontaneously experienced, including intrinsic values, cannot be denounced as less real than that of scientific theory, because we always ultimately refer back to immediate reality. . . . We are forced by modern science back to nature, basically as the naturalists conceived it. And it is, in its essential features, worth protecting for its own sake."[49] Those structures knowable to the human mind, which we have articulated and codified into science, philosophy, and religion, retain their qualitative richness only "if they continue conjunct and integral with this deep complex structure of quality built up through countless ages before even the human mind appeared and now accessible to the feeling-reactions of the human organism. But when the human mind in its pride tries to rear its knowable structures as supreme goals of human endeavor, impoverishment, destruction, conflict, and frustration begin because these structures are then cut off from the rich matrix of quality found in organic, nonintellectual reactions."[50]

Here, however, we must guard against our apparently inevitable tendency to fall into a surface ecoethic. We can neither justify the belief that creative evolution exists for us nor think that we are its goal and purpose. By realizing that we are part of cosmic process we confirm that we are finite beings bound in the reality of Becoming. We are not the privileged children fashioned in the image of God but coordinate interfaces of the historical process of nature. We do not impose value on a valueless cosmos; rather, we are sensitive registers of values created through the unfolding of time.[51]

SUMMING UP

I have argued that six aspects of the idea of wilderness offer deep ethical support for the conservation of biodiversity. *Practically considered*, we have neither the knowledge nor the economic capital necessary to replace historically evolved life systems. Manage we must, but to conserve life rather than to create a prosthetic environment. *Psychologically considered*, the idea of wilderness calls us back to the reality that we are of and about earth—not the masters and possessors of nature. *Politically considered*, the idea of wilderness leads us toward wild culture, a culture rooted in the particularity of place. Wild culture checks globalization. *Ethically considered*, the idea of wilderness reveals the anthropocentric bias of shallow or surface ethics and promotes a process of deep questioning that precedes any truly deep ecological ethic. *Ecologically considered*, the idea of wilderness calls our attention to the importance of knowledge of the deep structures that underlie and sustain any viable culture. *Cosmologically considered*, the idea of wilderness confirms human species as bearers of sentient witness to the reality of creative evolution.

Notes

1. Edward O. Wilson, *The Diversity of Life* (Cambridge: Harvard University Press, 1992), p. 32 and passim.
2. For an accessible discussion of wildlands conservation in North America

see "The Wildlands Project," *Wild Earth* (Special Issue), available through Wild Earth, P.O. Box 455, Richmond, VT 05477. For a more theoretical discussion of conservation biology see Michael E. Soulé, ed., *Conservation Biology: The Science of Scarcity and Diversity* (Sunderland, Mass.: Sinauer, 1986). Each continent, it must be noted, and each bioregion within each continent, faces unique challenges.

3. See, for example, William C. Clark, "Managing Planet Earth," *Scientific American* 261(3) (1989):46–54. The managing planet earth (MPE) strategy assumes that the economic needs of a growing human population can be met and that we can also preserve biodiversity through better planning, more efficient technology, and development. For a short critique of the MPE or high-impact path see John Firor, *The Changing Atmosphere: A Global Challenge* (New Haven: Yale University Press, 1990).

4. For a case study in the failure of the sustainable development ideology see Donald Ludwig, Ray Hilborn, and Carl Waters, "Uncertainty, Resource Exploitation, and Conservation: Lessons from History," *Science* 260 (1993):17. See also Nicholas Georgescu-Roegen, *The Entropy Law and the Economic Process* (Cambridge: Harvard University Press, 1971).

5. See, for example, Neil Evernden, "Ecology in Conservation and Conversation," in Max Oelschlaeger, ed., *After Earth Day: Continuing the Conservation Effort* (Denton: University of North Texas Press, 1992).

6. Soulé argues that "the only feasible way to maintain the vast majority of species is with very large, very secure nature reserves. Vest-pocket reserves and integrated agro-ecosystems cannot ultimately preserve more than a tiny fraction (probably much less than 10 percent) of the biota"; Soulé, *Conservation Biology*, p. 369.

7. Wilson, *Diversity of Life*, pp. 348–349.

8. See Max Oelschlaeger, *The Idea of Wilderness: From Prehistory to the Age of Ecology* (New Haven: Yale University Press, 1991), chap. 10, for a fuller account.

9. Although some critics claim that, given the relentless humanization of the planet, the notion of managing wilderness is a contradiction in terms, management of some kind is the sine qua non of conservation. The difficult issues concern the goals of management. See John C. Hendee, George H. Stankey, and Robert C. Lucas, *Wilderness Management*, 2nd ed. (Golden, Colo.: North American Press, 1990), for discussion.

10. A distinction between the knowledge required for the theory and practice of conservation biology and the knowledge required for the continuation of the historical project to manage planet Earth is apropos. The former is a *path of humility*: where research aims at articulating nature's semiosis (minimally involving population biology, biogeography, community ecology, evolutionary biology, and other biological and physical sciences) to

the end of preserving biodiversity. The latter is a *path of arrogance*: research aims at articulating a cultural semiosis (minimally involving biotechnology, ecotechnology, environmental engineering, and environmental economics) to the end of the continued exploitation of the planet as economic resource to sustain the life of one species—namely, our own. Given the end of preserving biodiveristy, there is an apparent consensus that management can proceed on the basis of imperfect knowledge. Yet many scientists are skeptical that the managing planet earth strategy can be successful, given the manifest insufficiencies of knowledge. John Firor argues in *Changing Atmosphere* (p. 103) that there are "many uncertainties" inherent in the project to manage the planet, since it "requires not only that we replace the evolved systems on which we depend, but also that we be wise enough to do so in the right sequence and completely, so that at no time during the process are we left without food to eat, air to breathe, and strong governments to keep the peace. This qualification is needed because the increasing domination of natural systems by our expanding technological society also means that in addition to freeing ourselves from our dependence on these systems, we are also causing their gradual disappearance. We will need to replace the accumulated 'wisdom' of the interconnections between air, earth, water, and species with our own intelligence, diligence, and management skills."

11. See Max Oelschlaeger, *Caring for Creation: An Ecumenical Approach to the Environmental Crisis* (New Haven: Yale University Press, 1994), especially pp. 1–83.

12. See Gilbert F. LaFreniere, *Ideas of Progress: A Collection of Primary and Secondary Readings* (Salem, Ore.: Willamette University Press, 1987).

13. See Firor, *Changing Atmosphere*, p. 103ff.

14. See Vandana Shiva, *Staying Alive: Women, Ecology, and Development in India* (London: Zed Books Limited, 1989).

15. See, for example, Firor, *Changing Atmosphere*, and Neil Evernden, *The Natural Alien: Humankind and Environment* (Toronto: University of Toronto Press, 1985). For multicultural perspectives see Shiva, *Staying Alive*, and Vine Deloria, Jr., *God Is Red* (New York: Dell, 1973).

16. Ecopsychology is a growing discipline. See, for example, Hans Peter Duerr, Claude Lévi-Strauss, Gregory Bateson, and James Tillman. See especially Paul Shepard, *Thinking Animals: Animals and the Development of Human Intelligence* (New York: Viking Press, 1978), and also *The Tender Carnivore and the Sacred Game* (New York: Scribner's, 1973). Shepard's work remains unsurpassed in originality and the thoroughness with which natural history informs his account of human psychology.

17. See René Descartes, *Rules for the Direction of Mind*, trans. Elizabeth S. Haldane and G. R. T. Ross, in Robert Maynard Hutchins, ed., *Great*

Books of the Western World, vol. 31 (Chicago: Encyclopedia Brittannica, 1952), p. 3. Also see Ernst Mayr, *The Growth of Biological Thought: Diversity, Evolution and Inheritance* (Cambridge: Harvard University Press, 1982), p. 79. Mayr provides an excellent discussion of the influence of the physical sciences on the shaping of the modern world, noting that "it was a tragedy both for biology and for mankind that the currently prevailing framework of our social and political ideals developed and was adopted when the thinking of [the Western world] . . . was largely dominated by the ideas of the scientific revolution, that is, by a set of ideas based on the principles of the physical sciences. . . . When evolutionary biology developed in the nineteenth century, it demonstrated the inapplicability of these physical principles to unique biological individuals, to heterogeneous populations, and to evolutionary systems." But the established socioeconomic system proceeds oblivious to the implications of evolutionary biology; the consequence is the anthropogenic mass extinction of life.

18. See Christopher Manes, "Nature and Silence," *Environmental Ethics* 14 (4)(1992):339–350. See also Richard Rorty, *Contingency, Irony, and Solidarity* (New York: Cambridge University Press, 1989), concerning the contingent nature of legitimating narratives and final vocabularies.

19. See Maurice Merleau-Ponty, *The Visible and the Invisible*, trans. A. Lingis (Evanston: Northwestern University Press, 1968), and Paul Ricoeur, *Main Trends in Philosophy* (New York: Holmes & Meier, 1979).

20. See Max Oelschlaeger, "Earth-Talk: Conservation and the Ecology of Language," unpublished paper presented at the Fifth World Wilderness Conference, Tromsö, Norway.

21. Ernest Becker, *The Denial of Death* (New York: Free Press, 1973). See also Max Oelschlaeger, "History, Ecology, and the Denial of Death," *Journal of Social Philosophy* 24(3)(1993):13–39.

22. Natural process itself is historical, whatever perspective (cosmological, physical, biological) is taken. However, our sense of human history is conventionally defined as outside of natural time. See Carl Friedrich von Weizsäcker, *The History of Nature*, trans. Fred D. Wieck (Chicago: University of Chicago Press, 1949), and J. T. Fraser, *Of Time, Passion, and Knowledge: Reflections on the Strategy of Existence*, 2nd. ed. (Princeton: Princeton University Press, 1990).

23. John Muir, *My First Summer* (Boston: Houghton Mifflin, 1911), p. 39.

24. Wilson, *Diversity of Life*, p. 272.

25. See, for example, Herman E. Daly, *Steady-State Economics*, 2nd. ed. (Washington, D.C.: Island Press, 1991), and Paul Elkins, Mayer Hillman, and Robert Hutchison, *The Gaia Atlas of Green Economics* (New York: Anchor Books, 1992).

26. Julian Burger, *The Gaia Atlas of First Peoples: A Future for the Indigenous World* (New York: Anchor Books, 1990), p. 78.

27. Shiva, *Staying Alive*, passim.

28. Enormous racial inequities also exist within the developed nations, such as the United States. See, for example, "Texas Environmental Equity and Justice Task Force Report" (Austin: Texas Water Commission, 1993).

29. Murray Bookchin, *Remaking Society: Pathways to a Green Future* (Boston: South End Press, 1990), p. 171.

30. See Richard B. Lee and Irven DeVore, eds., *Man the Hunter* (New York: Aldine De Gruyter, 1968). Judged historically, urban industrial civilization is the exception, not the rule.

31. See Gary Snyder, *The Practice of the Wild* (San Francisco: North Point Press, 1990).

32. See Daniel Kemmis, *Community and the Politics of Place* (Norman: University of Oklahoma Press, 1990).

33. Edward O. Wilson, *Biophilia* (Cambridge: Harvard University Press, 1984), pp. 138–139. Topophilia, the love of place, is another powerful psychic motivation. See Yi-Fu Tuan, *Topophilia: A Study of Environmental Perception, Attitudes, and Values* (Englewood Cliffs: Prentice-Hall, 1974).

34. See Oelschlaeger, *Idea of Wilderness*, especially chap. 2.

35. The land ethic has found its way into the wilderness management literature. See, for example, Richard B. Primack, *Essentials of Conservation Biology* (Sunderland, Mass.: Sinauer, 1993), p. 247, and Oliver S. Owen, *Natural Resource Conservation: An Ecological Approach*, 5th ed. (New York: Macmillan, 1988), p. 1.

36. Aldo Leopold, *A Sand County Almanac: With Essays on Conservation from Round River* (San Francisco: Sierra Club Books, 1970), p. 279.

37. See Will Wright, *Wild Knowledge: Science, Language, and Social Life in a Fragile Environment* (Minneapolis: University of Minnesota Press, 1992), Arne Naess, "How Should Supporters of the Deep Ecology Movement Behave in Order to Affect Society and Culture?" *Trumpeter* 10(3) (1993): 98–100, and Max Oelschlaeger, "Wilderness in Postmodern Context," unpublished paper presented at the 1992–1993 MacArthur Workshop on the Sources of the Emerging Environmental Consciousness, Massachusetts Institute of Technology, Cambridge, Mass.

38. Merleau-Ponty, *Visible and the Invisible*, p. 155.

39. Wright, *Wild Knowledge*, p. 114.

40. Jerry F. Franklin and Ed Bloedel, "Wilderness Ecosystems," in Hendee, Stankey, and Lucas, *Wilderness Management*, p. 243. See Soulé, *Conservation Biology*, pp. 233ff. for a technical discussion of the so-called surface effects of human population on biodiversity.

41. Adapted from Franklin and Bloedel, "Wilderness Ecosystems," pp. 250ff. and Soulé, *Conservation Biology*, passim.
42. See Eugene P. Odum, *Ecology and Our Endangered Life-Support Systems*, 2nd. ed. (Sunderland, Mass.: Sinauer, 1993).
43. See Snyder, *Practice of the Wild*.
44. Henry David Thoreau, "Walking," in *Excursions and Poems*, vol. 5 of *The Writings of Henry David Thoreau* (Boston: Houghton Mifflin, 1906), p. 240.
45. David Bohm, *Causality and Chance in Modern Physics* (Philadelphia: University of Pennsylvania Press, 1957), pp. 152–153.
46. See Ilya Prigogine and Isabelle Stengers, *Order Out of Chaos: Man's New Dialogue with Nature* (New York: Bantam Books, 1984), and Ilya Prigogine, *From Being to Becoming: Time and Complexity in the Physical Sciences* (New York: W. H. Freeman, 1980).
47. See John D. Barrow and Frank J. Tipler, *The Anthropic Cosmological Principle* (New York: Oxford University Press, 1986).
48. Henry Nelson Wieman, "The Source of Human Good," in Pete A. Y. Gunter and Jack R. Sibley, eds., *Process Philosophy: Basic Writings* (Washington, D.C.: University Press of America, 1978), p. 379.
49. Arne Naess, "Intrinsic Value: Will the Defenders of Nature Please Rise?" in Soulé, *Conservation Biology*, p. 505.
50. Wieman, "Human Good," p. 379.
51. See Samuel Alexander, *Space, Time and Deity* (Gloucester, Mass.: Peter Smith, 1979). According to Alexander, if divinity is outside cosmic process then God does not exist, since there can be no being apart from becoming.

Wilderness and Human Habitation

DAVID JOHNS

Wilderness has come under new attacks of a different sort over the last several years.[1] These attacks have come from within the environmental movement itself and are a reaction to a number of purported defects in wilderness thinking: opposing wild nature versus humans; failing to recognize the need to integrate people into nature; and failing to realize that people have always shaped nature. I want to address this set of criticisms.

Criticism One: *Wilderness exists only from the perspective of the paradigm of civilization. Wilderness reinforces that paradigm, rather than overcomes it.*

The idea of wilderness emerges out of the essential dualism that characterizes civilization and the thinking/emotional processes that are part of it. Civilization sets itself up as separate from and hostile to wild land—that is, land not under human control.[2] But wilderness is not merely a category in a paradigm or mental construct. There is a vast qualitative difference between lands that are exploited by civilization (and its hierarchical and usually sedentary precursors) and lands that are not. Wilderness is the term ecocentric or biocentric people give to land that has not been significantly degraded by humans, land that still supports ecological processes and indigenous biodiversity.

To stop thinking in terms of the wilderness/civilization dichotomy does not resolve the problem of civilized societies colonizing and consuming ecosystems, much as a malignant tumor consumes surrounding cells and tissues.[3] Human history and prehistory are rich with examples of those who diminished or destroyed what they claimed to respect. To transcend the dualism between civilization and wilderness we need to change behavior as well as thinking and feeling. While we work toward the material transcendence of dualism, we must protect what is left of the wild and restore what is needed to ensure the integrity of ecological processes.

Supporters of wilderness recognize that in the near and medium term, if not the long term, the essential nature of civilization is not likely to change, and life on this planet needs to be protected from it. This is not to say that we should abandon the goal of fundamentally altering the relationship between humans and the rest of nature. It is to recognize that we cannot wait for that change to protect biodiversity or bioevolutionary processes.

The term *wilderness* evolved from earlier Celtic words meaning "self-willed land."[4] Self-willed land is neither tame nor domesticated. It is land free from human colonization. (Colonization is not the mere presence of humans but the conversion of ecosystems to the dominant use by a single species and the resulting decline in diversity, complexity, and evolutionary dynamism.) Large numbers of people and high levels of consumption are invariably founded upon colonization. To protect landscapes as wilderness is to protect their self-willed character—to protect them from the overarching willfulness of a single species. To protect areas as wilderness is to protect biodiversity and the dominance of ecological processes. Not to protect wilderness in the name of transcending dualism is to leave relatively whole and healthy landscapes vulnerable to destruction.

Criticism Two: *People belong in nature, not apart from it. The problem with wilderness is it keeps people, including indigenous people, apart from nature.*

The separation of people from nature is indeed the problem. Bringing people back into nature is indeed the solution. Touching the hearts of people, helping them recover their deep links to the planet that nourishes them and reawakens their love and respect for the rest of life, is essential. It is not accidental that rites of passage in many precivilized societies involved going into the wilderness alone. There one found one's deepest self as a part of the living landscape. We risk, without wilderness, losing landscapes that make such encounters possible. If Paul Shepard is correct, in losing wilderness we risk losing any chance to overcome our arrested development.[5] In losing wilderness, we lose the world which gave us birth.

Reawakening such feelings is only a beginning—the starting point of a long road on which humans must change the way they live and reduce their numbers. How are people to be brought back to some deeper relationship with nonhuman nature? Many point to peoples who live simply as having important lessons to teach those caught up in the world's dominant cultures. Indeed, many who criticize wilderness point to the role that human groups play in enhancing biodiversity. We'll return to this point later, but first let us look at what we might learn from those who live close to the land.

Whether we assume groups of people having a close relationship with the earth know something we don't, or whether they simply lack the capacity to be as destructive, is beside the point here. When certain societies are described as living within ecological processes, such as the Kung (southern Africa), Inuit (northern Canada), or Kayapo (Amazonia), two things stand out that are anathema to civilized societies. The first characteristic is a combination of very low population and nonintrusive technology, neither of which should be equated with cultural simplicity. The second characteristic is a largely egalitarian society devoid of aggrandizing schemes of conquest, accumulation, or other sorts of domination.

When human numbers increase much beyond band size and concentrations, the relationship with nature changes because people must move toward more manipulative and intrusive technologies and forms of social organization. More of the ecosystem must be turned toward meeting human consumption, and there is less for other species. Controlling nature to support greater human numbers, in turn, leads to diminished feelings of kinship with nature.[6]

How are 5.5 billion people to get close to the land? How are 400 million North Americans to get close to the land? The 400 million North Americans now alive could determine to reduce their collective impact by drastically lowering levels of consumption, but current human numbers would continue to necessitate turning vast parts of the earth into tame gardens and farms. Our numbers, and the social organization that supports those numbers, lock us into a war with the wild. Our numbers make us poor. We react with intolerance toward other creatures that use the space or food we want. We must have it all.

During most of humankind's time on this planet we have been few in numbers, spreading slowly across the earth, with the limited technology and social organization necessary to support bands of gatherers and hunters. The last ten thousand years, especially the last three hundred, have seen that situation radically altered. We are no longer a few million. We no longer live by gathering supplemented with fishing, hunting, and scavenging, but by way of agriculture in all its permutations.

During most of human history we have not lived in large concentrations nor as settled populations. Until recently many areas were not occupied by humans, or were occupied only seasonally or transiently, and many areas were used only in a limited way—for sacred purposes, as buffers between groups, and so on.[7] Humans in band society are not unlike other mammals in terms of affecting the evolution of ecosystems. As human densities approach those requiring horticulture or agriculture (or some equivalent, like the salmon runs in the Pacific Northwest), our species' impact goes significantly beyond a coevolutionary one. While there are some examples of humans acting as a keystone species (like beavers) increasing diversity, the usual and historically overwhelming result is ecological disruption and degradation—notwithstanding the view of nature held by the people involved. Indeed, people's view of nature tends to follow their adaptive behavior, however alienating.[8] Even prior to agriculture and horticulture the adoption by some groups of big game hunting may have resulted in extinctions.[9]

Agriculture allowed rapid expansion of human numbers and accumulation of surpluses which provided room for changes in social organization, technology, and attitudes usually described as civilized. The emergence of civilization (cities, states, great hierarchies, professional armies) marks the adolescence of the human ability to colonize nature: to utterly bury ecosystems under cities and fields, to reach out to other ecosystems for materials, to replace diversity with monoculture, and to consume entire rivers for irrigation.

Large numbers of people, and the social/technological processes that go hand in hand with civilization, undermine ecological processes and destroy biodiversity. This is true of the Aztec, Inca, Chinese, Indian, and Sumerian civilizations, as well as the European. Large numbers of people and certain kinds of technology degrade landscapes. They destroy biodiversity. A friendlier attitude toward the earth and other life may lessen the impacts—compared to a society that reduces nature to resources in its thinking—but it will not fundamentally ameliorate them. The existence of competing centers of civilization greatly exacerbates the negative impact, because centers see other centers as threats to their hegemony and step up their attempts at dominion and extraction.

The argument that people have been around for a long time everywhere, having a significant impact on nonhuman nature, demonstrates an odd inability to discern the difference between 50 million humans and 5 billion and the difference between 4 billion years (the approximate time life has existed on earth) and the length of time *Homo* has been here. To claim that human beings, once they have achieved the ability to affect biodiversity, have on the whole improved it, smacks of the same dreary humanist arrogance that marks the worship of the noble savage or the Renaissance man.

Biodiversity and ecological processes need protection from these billions of humans—protected areas where people do not reside, except in small numbers with pre-Mesolithic technologies. It is simply not possible for 5 billion people, even living at the level of the poor in the Third World, to live on the earth without seriously impoverishing the biosphere. It is impossible for a billion people to live as North Americans, Europeans, or Japanese do, and not seriously degrade the biosphere, not to mention their fellow humans. Putting that number of people in touch with nature by allowing roads, large numbers of recreationists, or other artifacts of civilization in the few remaining areas where biological processes still have some wholeness, so that they may experience a conversion, would be a disaster. Not to restore areas—on these or other grounds—would also be a disaster.[10]

Criticism Three: *Indigenous people had a profound impact on nature—indeed they managed nature. Such stewardship is fundamental to human beings and important to the health of nature.*

Much depends on how one thinks about the term *indigenous people* and what constitutes management or stewardship. In the Americas, indigenous people refers to pre-Columbian peoples and their descendants. For some purposes it may be useful to think of all such people as a category, but in terms of the human relationship to nonhuman nature, more is obscured than revealed in doing so. Pre-Columbian America was home to a wide range of cultures from the Inuit to the Aztec to the many bands in Amazonia. The attitudes as well as behavior of different groups toward the earth varied greatly. The Aztec overpopulated and destroyed much of the biodiversity and carrying capacity of the Valley of Mexico and surrounding area. Swidden agriculturists were not disruptive in small numbers, but in larger numbers they altered the forest, changing microclimates and causing extinctions. In some instances burning led to erosion and long-term changes. Some agriculturists cut huge forests over the decades, reducing biodiversity and undermining the ecological basis of their society.[11] Anthropologists and paleontologists debate whether big game extinctions were caused by climatic shifts or intensive big game hunting. Most believe big game hunting played at least some role.[12]

Other groups of North American indigenous peoples surely did live in nature, as one species among many. But there is no monolithic Indigenous People that can serve as a model. There are many indigenous groups, including those that have adapted significantly to the modern world, and they have much to teach those of us whose indigenous roots are even more deeply buried than theirs.

If the term *indigenous* is not uniformly indicative of ecologically sound behavior, neither are the terms *management* or *stewardship*. All species change their environment. Large mammals have a significant impact on other species, both direct and indirect through their impact on habitats. As species evolve and expand (or contract) their range, ecosystems change. Harold J. Morowitz has suggested the real units of evolution on earth are ecosystems: a community of plants, animals, hydrological, trophic, and nutrient cycles. Stephen Jay Gould argues for a hierarchical view: genes, individuals, populations, species, and communities may all be subject to natural selection. Lynn K. Margulis and others have long argued the earth itself is best understood as a living entity with coevolving parts.[13] While evolution (or coevolution) does not have a purpose or goal, it does seem to have a direction. In a universe tending toward entropy some parts of the universe, receiving energy from a stable source and able to give it off to a sink at a correspondingly stable rate, tend toward greater complexity.[14] This has been the overall pattern on earth, five major extinction events notwithstanding.

What, then, is problematic about human behavior? Don't we simply coevolve in a manner similar to other large mammals, altering the overall whole? Sometimes we do; but we go beyond that. Our capacity for culture gives us the ability to break, at least temporarily, the solar economy—reducing complexity, diversity, and the integration of cycles and systems. Colonization, as opposed to coevolution, represents a cultural *choice* which involves reshaping, even destroying, ecosystems to serve human purposes.[15] Such reshaping has certain features: humans benefit to the detriment of other species and ecological processes generally; forests are transformed into tree farms or pasture, prairies into endless soybean fields, and rivers into poison soups, breaking down the capacity of the earth to cycle everything from nutrients to heat. Colonization diminishes the ability of landscapes and ecosystems to support diversity, replacing many species with a few or even one. Colonization means human numbers grow, sometimes rapidly, and consumption increases—all at the expense of other species. In short, self-regulation of the community as a whole is replaced by control of part of the whole, and spontaneity, liveliness, and biological integrity are diminished. The tiger is caged and oxen bred to the plow plod in rice paddies.

How do we distinguish this colonizing behavior from management or stewardship? The lines are not always clear, and the often polemical nature of the debate simply adds to the murkiness. I argue that the terms are so broad that they can be used to describe both colonizing and noncolonizing behavior. A forest is a living, self-regulating system. Humans, though they are recent arrivals, may, in limited numbers and with pre-Neolithic forms of social organization and technology, live as part of that system. Anthropogenic change may mimic nonanthropogenic forces or simply constitute one element among many.[16] Is this management, or does some other term apply? If it is in fact management, can the same term be used to describe people using (or suppressing) fire or engaging in other preindustrial practices that are ecologically harmful? Opponents of wilderness have argued that because pre-European North Americans managed their ecosystems and things were great up to 1492, we can do the same. Not only were things not uniformly great in 1492, but there is often a deliberate attempt to apply the term *management* to everything from bow and arrow to bulldozer. The Forest Service, BLM, and timber companies claim to be land managers and good stewards. Are we talking about simple misrepresentation or differences in scale? Did peoples with simpler technology have less impact, or did they have a fundamentally different relationship with nature? Or were some simply better managers than others? Are the goals simply different, or is there a fatal flaw in the entire notion of management? Doesn't it really mean human domination of one sort or another?

Clearly *management* is a very broad term. At issue, really, is the nature of human action with regard to nonhuman nature and whether such action constitutes a kind of colonization or constitutes living in and even restoring the landscape. What sorts of human intervention might be termed management that are also noncolonizing or ecocentric? The goals of human action must be considered in evaluating the propriety of management. Ecocentric human goals include restoring the conditions for an ecosystem's self-regulation and spontaneity. By removing degrading influences (industry, roads, pollution, exotic species, logging, dams, and so on) we may enable an ecosystem to begin healing itself, to reestablish successional patterns and resilience to natural disturbance regimes. Whether we call it management or not, restoring an ecosystem to self-regulation is very different from attempting to dominate it.

The goals of resource management, by contrast, are to maximize commodity extraction from an ecosystem—replacing natural relationships and processes with human-imposed ones. Such an effort involves ongoing inputs of energy to overcome unwanted processes and species and maintain exotics or preferred species, often monocultures. This sort of management is about control. It is directly opposed to self-regulation and spontaneity. In the end the difference between the two is the difference between power and love. The act of loving—allowing the

loved, the earth and other living things, to develop in their own ways—
is not an act of self-negation but a recognition that the conditions for
one's own unfolding are antithetical to power and can only occur in the
context of an integrated whole.

Stewardship commonly connotes benevolent management, but this
term too can be used to cover a multitude of sins. Paul Shepard observes
that many precivilized mythologies recognize and honor the evolution,
or in mythological terms the *unfolding*, of the larger whole. Humans
have a place in this whole, but not as lords of creation. It is only in civilized mythologies—which are in fact reified and dying mythologies—
that humans appear as lords, stewards, and managers.

Ecocentrism involves the recognition that we can only try to mimic
nature and must be cautious and respectful in our interactions, even in
attempting to assist nature in healing. Ecocentrists recognize that protected wilderness is important because it has worked for four billion
years and doesn't need to be reinvented. As Reed Noss has said, "Wilderness provides a standard of healthy, intact, relatively unmodified
land."[17] For wilderness to do this, it needs protection from industrial
civilization. Noss points out that areas set aside from people are the only
places large carnivores and other animals can exist in large, healthy
populations. Their health is indicative of the health of the system. Another definition of wilderness is the place of wild beasts; when the wild
beasts decline or disappear, the earth is poorer.

Ecocentric protection of wilderness does not mean we must walk
without shadows or footprints. It does mean living within certain limits: in terms of human numbers, technology, and levels of consumption. Wilderness means a recognition that evolution is wiser than any
single species, including us, and we must therefore live with humility.
It means abandoning our control fantasies for a reality within the bios-
pheric community. It means coming to terms with the fear of nature
within us as well as without, which is the motivation for control.

Wilderness is about learning how to live in the current—which may
entail swimming upstream at times—and not resorting to dams which
kill rivers. Wilderness is about living with wolves and bears and sharing the planet with them on equal terms.

The call for wilderness is not antihuman but a call for protection of the place that has long been our home. Wilderness is about preserving and restoring "self-willed" land and halting colonization. Wilderness must be protected from humans because all civilized societies, as well as many of their precursors, disrupt ecological processes and reduce biodiversity. Wilderness is needed to counter the lethal effects on biodiversity by large human numbers and certain kinds of social organization and technology. Until human societies can be radically altered, large portions of the earth must be protected from human intervention.

If I have offered more questions than answers it is because I have more questions than answers. Our work now is to get the questions right. Although it may sound from these arguments that I think we must go backward rather than forward, there is no going back in any simple sense. But if we look to human psychological healing, we may find a useful analogue. To heal a deep injury is to revisit it, reexperience it from an insightful perspective, and then move on along a different path. We need to heal the split. By injuring the earth we have injured ourselves. While we are caught up in the kind of violence that emerges from deep injury, we must be restrained. That is the purpose of wilderness protection and restoration. It is a stopgap until we learn to base our lives more on love than fear.

Notes

1. By wilderness I do not mean legal wilderness or some similar status. I concur with The Wildlands Project mission statement: "By wilderness we mean extensive areas of native vegetation in various successional stages, off limits to human exploitation; viable, self-reproducing, genetically diverse populations of all native plant and animal species, including large predators; and vast landscapes without roads, dams, motorized vehicles, powerlines, overflights or other artifacts of civilization, where evolutionary and ecological process can continue. Such wilderness is absolutely essential to the comprehensive maintenance of biodiversity. It is not a solution to every ecological problem, but without it the planet will sink into further biological poverty."

2. Some civilizations are ambiguous toward wilderness. Although they are generally hostile, they recognize it as a source of revitalization. See, for example, Paul Shepard, *Nature and Madness* (San Francisco: Sierra Club Books, 1982), and Peter Duerr, *Dreamtime* (London: Blackwell, 1985). In any event the civilized/wilderness dualism (as opposed to the distinction

between areas used for living, hunting, the sacred, and buffers from other groups) is central to the cosmology and emotional makeup of civilizations.

3. For purposes of simplicity I say civilization, but I also mean precivilized societies that are precursory to civilization, such as chieftaincies. See, for example, Elman Service, *Primitive Social Organization* (New York: Random House, 1962); *Origins of the State* (Philadelphia: ISHI, 1978); Ronald Cohen, *Origins of the State and Civilization* (New York: Norton, 1975); Robert McCormick Adams, *Evolution of Urban Society* (Chicago: Aldine, 1966); Hans J. Nissen, *The Early History of the Ancient Near East* (Chicago: University of Chicago Press, 1989); William Sanders and Barbara Price, *Mesoamerica* (New York: Random House, 1968).

4. Jay Hansford C. Vest, "Will of the Land," *Environmental Review* (Winter 1985):321–329.

5. See Shepard, *Nature and Madness*, and Morris Berman, *Coming to Our Senses* (New York: Simon & Schuster, 1989).

6. Mark Nathan Cohen, *The Food Crisis in Prehistory* (New Haven: Yale University Press, 1977).

7. On the evolution of our genus and its spread across the globe see Richard Klein, *The Human Career* (Chicago: University of Chicago Press, 1989); for an interesting interpretation see Jared Diamond, *The Third Chimpanzee* (New York: HarperCollins, 1992), especially pt. 4. See also Marvin Harris and Eric B. Ross, *Death, Sex, and Fertility* (New York: Columbia University Press, 1987). Of interest with regard to North America is William M. Denevan, *The Native Population of the Americas in 1492*, 2nd ed. (Madison: University of Wisconsin Press, 1992). See also John R. McNeill, "Agriculture, Forests, and Ecological History," *Environmental Review* (Summer 1986):122–133; and Roy Ellen, *Environment, Subsistence and System* (Cambridge: Cambridge University Press, 1982).

8. See Morris Freilich, *The Meaning of Culture* (Lexington, Mass.: Xerox College Publishing, 1971); and Marvin Harris, *Cultural Materialism* (New York: Random House, 1979). See also Marshall Sahlins, *Culture and Practical Reason* (Chicago: University of Chicago Press, 1976).

9. Paul S. Martin, "Prehistoric Overkill: The Global Model," and Richard G. Klein, "Mammalian Extinctions and Stone Age People," both in Paul S. Martin and Richard J. Klein, *Quaternary Extinctions* (Tucson: University of Arizona Press, 1984).

10. Even though wild areas are influenced by anthropogenic change elsewhere on earth, they are better than no wild places; see Reed Noss, *Conservation Biology* 1 (March 1991):120. The degradation of wild places by distant anthropogenic change is not a sound argument against wild places, but rather an argument for more and bigger wild places (they should be the matrix) and for limiting certain human activities wherever they occur.

11. See, for example, William T. Sanders and Robert S. Santley, *The Basin of Mexico* (New York: Academic Press, 1979), Roy Ellen, *Environment, Subsistence, and System*, and Andrew Goudie, *The Human Impact on the Natural Environment*, 3rd ed. (Cambridge: MIT Press, 1990).

12. See Martin and Klein, *Quaternary Extinctions*.

13. Harold J. Morowitz, *Energy Flow in Biology* (New York: Academic Press, 1968); Stephen J. Gould, "Darwinism and the Expansion of Evolutionary Theory," *Science* 216 (April 1982):380–387; Lynn K. Margulis, *Symbiosis in Cell Evolution* (San Francisco: Freeman, 1981); Lynn K. Margulis and Dorian Sagan, *Origins of Sex* (New Haven: Yale University Press, 1986); and Lynn K. Margulis, Mitchell Rambler, and Rene Fester, eds., *Global Ecology* (Boston: Academic Press, 1989).

14. See Morowitz, *Energy Flow*.

15. By choice I mean that human culture can take a variety of forms; I do not mean to imply here consciousness of alternatives or forethought.

16. See the discussion by Holmes Rolston III, "The Wilderness Idea Reaffirmed," *Environmental Professional* 13 (1991):370–377; and Reed Noss, "On Characterizing Presettlement Vegetation: How and Why," *Natural Areas Journal* 1 (1985):12–13.

17. Reed Noss, *Conservation Biology* 1 (March 1991):120.

The Cultured Wild and the Limits of Wilderness

PAUL FAULSTICH

The mind is a part of the nature of things, the world is a divine dream.
RALPH WALDO EMERSON, *Nature*

Let me make it clear from the outset: my vision for Earth is a planet populated sparsely by humans, where nonhuman processes prevail, and where alienation between humans and nature is absent.

Nature is a human construct imposed on the world, and we constitute it in ways that make cultural sense. It is an ideologically charged project, and knowledge about it is necessarily a social product. Nature is both something we observe and something we believe. It isn't simply an assemblage of things—as one sees in a natural history museum— but a way of understanding the world.

In order to comprehend more fully human relationships with the natural world, we should ask not just what nature means, but *how* nature means. Nature, I believe, is a condition of culture, not some sort of transcendental truth. It exists, in part, as an accumulation of human insights. It is easy to misinterpret this point. Nature is not simply something we dream up; no mind-is-all solipsism is intended. What humans perceive is not a product of our imagination. Rather, by attaching meaning to things, we reconstruct them and steep them in metaphor, thereby enabling ourselves to make sense of the world. Worldview and nature are closely related: worldview is constructed out of the salient elements of nature. But beyond all our ideas of it, nature is, of course, much more than we will ever know.

NATURE AND ECOLOGY

Ecology has emerged as the dominant philosophy informing contemporary perspectives on nature and the environment. Indeed, it is a political force which environmentalists, ecologists, and others use to promote their ideologies and agendas. Proponents of ecology have positioned themselves as authorities and claim their view of nature to be the most authentic. But Western paradigms of nature, from which ecology emerges, may be antithetical to holistic visions of a world remade. Ecology, for example, is mostly irrelevant to aboriginal peoples, who have their own legitimate views of how the world works. While I do not question the essential nature of ecology, I do call for an epistemological shift in the way we do business. The idea that there is, or could, or should be, a universally accepted ecology is merely another variant of habitual European appropriation.

In the face of global cultural diversity, the presumption that there is an authentic and systematic ecology bespeaks ecological imperialism. It implies an unachievable coherence and a grossly simplistic integration: an unnatural act. While ecologists have occasionally appropriated (and trivialized and misrepresented) indigenous philosophies, ecological insights rarely have been seized upon by Native peoples. Ecologists readily acknowledge that indigenous peoples have much to teach us, but we seldom demonstrate concern for what we can give back to the people from whom we've taken much. The ecological idea of Big Wilderness can be evoked both for and against the empowerment of indigenous peoples. Indeed, some of our most fervent endeavors to preserve wilderness have ended up thwarting indigenous efforts to maintain their subsistence and enhance empowerment. If, for example, indigenous people accepted the idea of wilderness as ecologists commonly render it (void of human habitation), it would destroy the indigenous presence.

Some ecological thinkers advocate reinventing our philosophies toward a more holistic approach to ecology and the environment. To this end we have borrowed generously from Native philosophies, but only when they happen to uphold the dominant scientific view of nature. Many indigenous peoples equate this kind of cultural appropriation with confiscating tribal lands or taking water rights. They feel they are being patronized and robbed. Native worldviews are among the most precious things that many indigenous peoples have to maintain their unique identities.

Efforts to address questions of ecological authority and appropriation have been limited, for the most part, to presentations of indigenous voices in ecological treatises. In all but a few instances, the approach has been to identify examples of indigenous philosophies that meet predefined criteria. The cultural bias of the Western ecological perspective has reinforced the neocolonial mythology surrounding contemporary indigenous people—that of the vanishing noble primitive living a simple and harmonious life. Strong and gentle natives at peace with nature; this has been the environmentalist's primitive. And while environmentalists may sympathize with the primitive's version of ecology, it is to the cultural lens of Western science that they inevitably turn for authoritative support for their visions. The mainstream scientific paradigm, however, continues to supply the rationale for mechanizing and appropriating the natural world. In an awkward embrace of this paradigm, environmentalists are bedding down with strange partners—and waking up with a dispassionate ecology.

ABORIGINAL KNOWLEDGE

How should nature be rendered? Whose version is to be accepted as authority? While it is largely the Western urban populations that determine international conservation policies and environmental priorities, the past decade has witnessed a shift in power relationships. This shift reflects the response of the environmental movement to broader ecological issues—in particular, the relevance of humans to wildlands. The growing visibility of indigenous peoples is especially important, since their ecological insights are, in many ways, different from the insights of other segments of the global community. Warlpiri Aborigines of Australia, for example, have an acute sense of bioregionalism in which people, spirit-beings, natural species, and geographic places are viewed as interconnected. For Warlpiri, nature is not about self, as distinct from other, but about the continuity of self and circumstance. This perspective differs significantly from that of industrial peoples, who speak of nature in terms of resources and regard it as something distinctly different from us. Warlpiri have no word for an exclusive nature.

Banjo Jungarrayi, an Aboriginal elder of Warlpiri descent, nostalgically recalls his adolescence: "We ran around naked," he says, "like wild men." Sitting on parched red sand amid flies and dogs, Jungarrayi recalls with regret his forced assimilation into a society where wildness is distinctly other than humanness. Despite fervent attempts by dominant Australian culture to discredit traditional Aboriginal life, Jungarrayi holds no shame for his youth. With pensiveness, Jungarrayi told me how his youth came to an abrupt end when he was chased down by men in a truck, lassoed, caged, and removed from his homeland. Now an old man, Jungarrayi wants nothing more than to return to Kunajarrayi, his "country," to once again be a wild man.

Jungarrayi's wildness does not require nakedness. What he desires is to live "out bush" where he can live according to the eternal myth of Jukurrpa—the Dreaming. Despite having been forcibly removed from his country at an early age and placed on government settlements, Jungarrayi and other Warlpiri have retained a strong relation with the land. Recent efforts to establish small breakaway communities—what Jungarrayi wants at Kunajarrayi—reflects the desire of many Aboriginals to renew long-standing social and symbolic attachments to the land. Known as "outstations," these communities are, in effect, homeland places, and the outstation movement reflects the desire of people to live more traditional lives closer to their traditional countries.

Aboriginals establish outstations to care for the country—to "look after 'im properly," as Warlpiri say. Although they have maintained traditional knowledge of totemic geography, many Warlpiri feel that by living at government settlements they are losing contact with important ceremonies and sacred places. Homesickness and a feeling of alienation from the land provide impetus for the move to outstations. This land-based renaissance has brought renewed use of traditional concepts of sacred sites and stewardship.

The outstation movement began, in part, as a way for Aboriginals to gain more political autonomy. Outstation living reduces Warlpiri participation in the dominant society; self-determination and self-image are thereby enhanced. As an assertion of independence, the revivalist outstation movement has enabled the reemergence of Aboriginal authority in spiritual, political, and ecological matters. Outstations typically consist of a few families who share an emotional attachment to the country to which they have returned. Although outstations do not represent an attempt by Aboriginals to return to a wholly traditional pre-European foraging existence, the movement does indicate a break from mainstream Australian traditions and values.

Jungarrayi's country, Kunajarrayi, is an imposing complex of sacred sites—an area pregnant with ancestral power that consists of numerous physiographic features that are transformations from the Dreaming: a waterhole was created where one of the Wolf Spider Ancestors emerged after traveling great distances underground; a long dike on a horizontal exposure of bedrock shows where a Witchetty Grub Ancestor turned itself into a Snake Ancestor after a fight and fled; granitic boulders are the transformed bodies of Ancestral Women that were raped and killed.

Aboriginal knowledge of Kunajarrayi is profound but not unique; Warlpiri can tell stories about every square meter of this desert. They know that if the cows are removed and the people return, the land will be whole again. Warlpiri are able to look after this land like no one else: they are inclined to "clean up the country" by lighting brushfires, and they perform the rituals that indicate they inhabit the land with respect. Respect for the land and its inhabitants is central to Aboriginal bioregionalism. This became evident to me during my early days in Warlpiri country, when an elder refused to shoot a wallaby, the animal of his Dreaming, that we encountered at a sacred site. "Him bin alright," I was told, "him bin sit down here longtime. Longtime fella that one."

Warlpiri maintain that the land both creates and is created by people. This notion has penetrating ramifications; persons and place share an identity. Nature is a repository of the most profound and recondite aspects of Aboriginal existence, and it holds the potency of the Dreaming. Old people may weep when they visit their homeland; they realize the full equation between country and culture. Aboriginal people returning to their homelands help dismantle the colonial histories which enabled Western societies to subjugate not only the people but also the land.

WILDERNESS AS CULTURAL CONSTRUCTION

Wilderness does not exist outside the human mind. Like concepts of nature, wilderness is a cultural construct; it exists as a paradigm, but not as an autonomous reality. Before the arrival of Europeans in the Americas, Native peoples participated in local ecologies, practicing controlled burning, seed propagation, and wildlife management. What was perceived by early European explorers as wilderness was already a humanized landscape. It is this superficially pristine environment—a wilderness that never existed—that we attempt to preserve in our national parks and reserves. Hence the "unpeopled wilderness" ideal of modern parks is substantially different from the wild inhabited lands actually encountered by the first European settlers. Established national parks, then, are based on the perverse fallacy that a wilderness unaltered by humans predated the European arrival. Interpretive centers in the parks present the prior aboriginal inhabitants as though they are bygone fauna. By underestimating anthropogenic interactions with American ecosystems during pre-Columbian times, we manage parks under the rubric of uninhabited nature—and have done as much to create a wilderness landscape as we have done to preserve it.

Wilderness reserves are not anomalous places liberated from human domination. Nature is not set free in these contexts. Rather, it is severely managed to maintain a comfortable image of wildness. In creating and managing such reserves we concurrently limit human use and enhance human domination. Ecosystems are appropriated in the name of environmentalism, and a puritanical vision of ecology is forced onto nature through a plethora of artificial management schemes. The process of identifying land worthy of preservation, of controlling land use by legislation, and of imposing boundaries on wildlands is entirely a cultural process. While this cultural process is not problematic per se (all peoples engage in such praxis), the dominant version of wilderness is dysfunctional and alienating. Wilderness reserves are the expression of an adversarial relationship in which wildness is subdued and controlled.

Wilderness areas or reservations represent an imperial policy of assimilation wherein nature is subjugated and given the "right" to exist as a colony, dependent upon the goodwill, management, and whim of a controlling body. Nature is expected, therein, to function in accordance with a series of laws and imposed schemata. Anthropomorphized through human values and ethics, it is expected to behave as a marginalized entity of civilized culture, enjoying the same rights and privileges, accepting the same responsibilities, and influenced by the beliefs of the dominant human community. Wilderness is patronized as an otherness and is effectively cordoned off from human cultures.

THE WILDLANDS PROJECT

Even enlightened models of wilderness remain exclusive. Consider one of the criteria of wilderness as defined by The Wildlands Project's mission statement: "extensive areas of native vegetation in various successional stages off limits to human exploitation." As this criterion suggests, it seems almost impossible for us to separate the destructiveness of post-Pleistocene human activity from the possibilities of appropriate human involvement with wildness. The difficulty comes when we try to discuss our experience of nature without recourse to a jargon which is the property of a destructive enterprise. Perhaps I am just stumbling into some sort of semantic pothole, but as a human ecologist I regard exploitation as any type of use, not just the destructive extraction of resources. Collecting a bundle of sedge with which to weave a basket is an exploitive act. Just because humans have succeeded in destroying most wilderness does not mean that utilizing nature is ethically reprehensible. Indeed, any efforts to save wildness must incorporate use. Exploitation is an ugly word, but it's not necessarily an ugly act; there are those who maximize exploitation, and there are those who optimize it.

The mission statement continues: "We recognize that most of Earth has been colonized by humans only in the last several thousand years." Don't tell this to Warlpiri in Australia, to Hopi in the American Southwest, or to !Kung San in southern Africa. They have very different knowledge of their origins and histories. Apart from those circulating and breathing wildernesses known as oceans, most of what we call wilderness has been traversed or occupied by *Homo sapiens* for eons. The established ecological paradigm can be challenged on its capacity to fossilize the past and perpetuate the dubious opposition between people and wildness.

My intent is not to derail efforts to create expansive wild areas where biodiversity and nonhuman ecological processes dominate. Indeed, The Wildlands Project's mission statement was drafted by some of the most far-reaching minds in the modern conservation movement. But this only shows how slippery the idea of wilderness is. We need to reframe our understanding of the relationship between humans and nature—to discard this model of wilderness and replace it with the more organic concept of wildness. Wilderness separates; wildness integrates. Wilderness is a place; wildness is a condition. Humans are wild too.

WILDERNESS PRESERVATION: BATEK NEGRITOS OF MALAYSIA

While doing research in the tropical rainforest of Malaysia my wife and I went hunting with a Batek Negrito man named Kumbang. After walking for a time through the steamy jungle on narrow Batek trails, we heard a rustle in the canopy that signaled what we were after—gibbon. Slowing our movements to a meditative pace, we turned and faced the animal. All we could see was a tiny patch of dark hair: the bulk of the gibbon was obscured by foliage. Kumbang slipped a dart into his blowpipe, took careful aim, and fired into the canopy. *Phhfffft.* Silence. The dart missed. On the next attempt Kumbang pierced the gibbon's flesh with a poisoned dart. Branches exploded as the injured animal fled. Our hunting partner slipped off his Nikes and took off through the understory on an hour-long chase that ultimately afforded nothing.

Despite the Nikes, this was an experience more wild, more organic, than most backcountry immersions I have had in unpeopled wilderness. This is not an attempt at romancing the stone age; indeed, Batek are opportunists and have incorporated aspects of the outside world into their own culture, as is evidenced by the Nikes. Traditional Batek territory lies within the boundaries of Taman Negara, Malaysia's principal national park, and Batek foraging activities (which include collecting jungle produce for trade) do not conform to park regulations. Government policies, pressuring them to leave their ancestral homeland in the name of wilderness preservation, undermine Batek autonomy and lifestyles. Despite their resistance, Batek have been relocated outside the park boundaries. At their new settlement Batek are encouraged to emulate Malay subsistence farming communities, as their seminomadic existence is perceived by authorities as primitive. But Batek are hunters and they are gatherers; they have no desire to be sedentary farmers. They continually transgress park regulations by pursuing their seminomadic culture within the park boundaries.

Taman Negara was established to preserve the flora and fauna of Malaysia's equatorial jungle. The protection of this indigenous biota has not taken into full account the Batek, however, whose culture is an integral part of the rainforest ecology as we know it. For example, Batek advertently and inadvertently distribute the seeds of rainforest fruits throughout their territory. This, in turn, influences the demographics of animal species and, in conjunction with traditional hunting practices, helps maintain faunal population balances. Denying Batek full access to the land, therefore, may preclude the continuation of cultural practices which are consequential to rainforest ecology.

Beyond ecological justification of their presence in the rainforest, however, Batek know this land as their ancestral homeland. This, as much as any scientific justification, is why they should remain in their territory. Blind adherence to the national park and wilderness ideals can be disruptive to both native peoples and natural systems. Diversity is fundamental to ecology, and biological diversity might very well be linked to cultural diversity.

Wilderness preservation vis-à-vis the exclusion of people backfires in another way: by precluding appropriate local use, we place even greater demands on global exploitation. Not only are we moving the problem somewhere else, but we are increasing both energy consumption and dependence on multinational corporations. The difference between environmental adaptation and environmental exploitation has been highlighted by Raymond Dasmann in a distinction he makes between "ecosystem" and "biosphere" peoples. Ecosystem peoples have developed small-scale societies and depend on resources supplied by local ecosystems. They know almost immediately if their exploitation patterns are damaging. Biosphere peoples are industrialists and extract resources from throughout the globe. They (we!) may not be aware of, immediately affected by, or even particularly concerned about the destruction of distant ecosystems. By exporting our demands, we pretend to escape the ecological constraints of local ecosystems, and we become increasingly unresponsive to natural equilibrium mechanisms.

ECOLOGICAL COLONIALISM

Indigenous peoples have no need for wilderness. To many, the wilderness ideal has come to represent just another expression of exclusionary colonialism. Wilderness is one more resource—this one being exploited by environmentalists (along with backpackers, biologists, biotechnology firms, and tour companies). Over and again, the idea of wilderness has been used to keep people—especially Native people—away from an area, such as the notion expressed in the Wilderness Act of 1964 that wilderness is a place "where man himself is a visitor." The popular notion of the national park in North America is that of an unpeopled sanctuary in which to escape. But our relationship with nature is not so simple, and human habitation need not be disruptive of natural systems.

What we have created with the wilderness and national park ideals is an artificial reality in which synergistic human/nature relations are excluded while dysfunctional relations (sightseeing, RV camping, shopping for souvenirs) are encouraged. Our experience of nature is shaped by the rhetorical constructs of culture in which we have taken it to be our beneficent burden to incarcerate wilderness. By recognizing the neocolonial uses of ecology we can begin to see the part that the Western environmental movement has played in creating a destructive, rather than instructive, model of nature.

We have tricked ourselves into believing there is a single ecological reality that aligns with our Western cultural predilections. In so doing, we subvert the diverse and organic structure of nature. This is not to suggest that the spiritual and intellectual roots of modern environmentalism are narrow; indeed, they are spread wide. But in our quest to understand nature we continue to reduce it to things. This perspective simplifies nature and is used to justify our rampant exploitation of it. In recent decades we have understood nature as interrelationships between things, but ultimately it is still "stuff" which is at the center of our ecological paradigm. As it has been mainstreamed, ecology supports maximum utilization (which may include preservation) of the material world. Our idea of nature, therefore, is embedded in the very obsession that has enabled us to destroy it.

Colonialism has not abated; it continues with intensity. The players, however, are different. Developers, corporations, governments, banks, real estate speculators, and, yes, ecologists are among today's colonialists. As a force that seeks to bring divergent values and perspectives into one singular model that is authoritative, colonialism is evident in ecological thought and Western environmentalism. Today ecology provides colonialism with an imperative justifying the conquest of nature: in its compromised expression, it creates dualities that allow some areas to be "sacrificed" in the name of progress while other areas (usually tiny in comparison) are "preserved" as wilderness.

Ecological colonialism arises from power relationships affecting the ways in which some versions of ecology are mainstreamed while others are relegated to the margins. Power in ecology is the ability to persuade people to support your version of nature. In its extreme form, dispassionate ecology is the parasitic McDonald's of the ecological paradigm: metastasizing all over the world, smothering local differences, packaged and shipped globally as *the* solution to saving imperiled ecosystems.

To what extent should human presence in wildness be welcomed? In our convoluted, industrialized world, some compromises may be necessary. Protected lands on which endangered species live, for example, should not be open to human access at all—unless, of course, symbiosis exists between threatened species and particular cultural patterns. The traditional burning of grasslands by Indians in northern Alberta, for instance, increases bison populations which at present are artificially low. (There are numerous other examples of symbiotic relations between cultural practices and wild nature.) Endangered species inhabiting unpeopled wildlife reserves should be regarded as in retreat—a retreat from which they may once again emerge in a healthy relationship with people.

I do not share the view that Native peoples are innately conservation-inclined. To a significant extent, it is available technology—be it stone tools or chainsaws—that sets limits on ecological exploitation. Traditional ecological knowledge is the basis of much cultural behavior that results in conservation practices, but the context and motivation for it are often different from that of Western resource management. Whether conservation is intended or not is perhaps irrelevant, since it is not possible to divorce the ecological aspects of a tradition from the technical, religious, or social. Native American creation myths, for example, convey not just knowledge of the environment but also the moral precepts which guide ecological practices. Thus worldview and ecology are one.

Attempts are now being made to design more holistic approaches to conservation, accommodating the lifestyles of Native peoples while preserving biodiversity. In Papua New Guinea, for example, traditional resource management practices are incorporated into the structure of wildlife management areas. These preserves are intended to provide an acceptable means of nature conservation while recognizing a local people's access to land. Participating communities abide by regulations which are based partly on such cultural considerations as land tenure and customary hunting rights. Communities may adopt nontraditional regulations as well, including bans on hunting or logging in designated areas. The importance of this approach is that it does not alienate people or wildlands.

The empowerment of indigenous peoples—and the way in which interaction between indigenous peoples, ecologists, and governments can alter ecological practices—has become a central concern of the politics of ecology. The issue should not be one of control, however, but one of developing partnerships based on the premise that there are multiple perspectives of nature (each equally valid) and multiple stakeholders. This notion becomes complicated only if it is assumed, as imperialism does, that what ultimately needs to emerge is a single correct ecology. Singular understandings of nature always result in exclusion and marginalization—most often of Native peoples since they hold the least power. Singular views of nature also undermine the fundamental ecological concept of interrelationship. In addition to sharing their own vision of nature, ecologists should seek to bring together diverse insights about global environments. Our limited understanding of nature can be transformed as different kinds of knowledge work together, and ecologists should not strive to be the final arbiters in creating, describing, and saving the world.

AN ECOLOGY WITHOUT BORDERS

Peoples have long played integral roles in the unfolding of natural systems. The small-scale and gradual environmental modifications of adaptive cultures are akin to ecological processes. The modifications caused by industrial peoples, by contrast, are global and rapid. While adaptive cultures live bioregionally and attune themselves accordingly, industrial cultures have, without exception, created exceedingly unstable and uniform ecological conditions. Nature exists as a self-regulating system, and throughout most of our cultural unfoldings *Homo sapiens* has participated in this system without destabilizing it. The line should be drawn, I think, not between humans and wilderness but between industrial processes and wild nature.

Insular wilderness is wildness which has been set apart from us and is antithetical to nature. How, then, is nature to be retold, the misrepresentations re-presented, the knowledge restored? We need people inhabiting the wild—reinventing a constellation of cultures where nature provides the model for thought and acton. As an inventive explication of nature, mythology can help us reposition ourselves in a synergistic relationship with the natural world. Mythology can be a magnificent blending of natural history and poetic speculation. We may think of nature as externalizing a construction, but it is more true, I think, to say that mythological constructions are our inward occasions of nature's presence. Mythologies are not indulgent dreamings but organic metaphors: artifices of the natural. The functional outcome of myth is not that it objectifies the natural but that it expresses human insights about our relations with nature. It engenders respect and a sense of belonging.

And this brings me to the proper business of ecology: is it ideas or things? It is, and it must be, both. Ultimately, things, as we know them, *are* ideas. Of necessity, then, ecology must be both a science and an art.

If we recover a narrative way of knowing, a knowing that has been devalued in scientific thought, we can begin to heal our alienation from wildness. The vital, sustaining images of mythology can immortalize our construction of wild nature, commensurate with our experience of it as wild humans. What will emerge is an ecology without borders and the understanding that there is more than one kind of knowing. Ecology, at its best, can help us live in a mythic world of ordered relationships where every human action becomes integrated into the ecosystemic order of nature.

Nature is a constellation of phenomena. Ideas of it vary according to the perspective from which these phenomena are perceived. Like stellar constellations, nature exists from the perspective of specific predilections and vantage points: Ursa Major looks like a big dipper to some of us; to others it appears as a great bear. What we see is a product of our heritage. Moreover, Ursa Major presents itself to us in the form that it does because of our earthly perspective. Viewed from outside our solar system, it might not appear as a coherent stellar assemblage at all.

Western ecological thinking provides no recipe for global environmental protection; nature is too marvelous and complex for monocultural paradigms. To be effective at saving and preserving wildness globally, diverse perspectives of nature must be incorporated. This culturized nature—the cultured wild—will enable us to break away from our frightened standoff with the world we inhabit. None of our disordered relations with the natural will change until we alter relations of power and accept nature as a model of diversity. What we must achieve ultimately, and soon, is self-determination of the wild. Toward this end, our environmental goal should be to seek organic models of nature and wildness that are inclusive rather than exclusive. We should strive to bring our ways of knowing into closer harmony with our ways of being wild-in-the-world.

To Warlpiri Aborigines, the Milky Way is the great Rainbow Serpent stretching luminously across the night. What they see is only natural.

The Quality of Wildness: Preservation, Control, and Freedom

JACK TURNER

Thoreau began talking about wildness as the preservation of the world in a lecture he gave at the Concord Lyceum on 23 April 1851 entitled "The Wild." In June of the following year he combined it with another lecture on walking and published the two as the essay "Walking, or the Wild" in *Atlantic Monthly*. This essay remains the most radical document in the history of the conservation ethic—and "how we understand that ethic depends on what we think Thoreau meant by *wildness*."[1]

Thoreau understood wildness as a quality: wild nature, wild men, wild friends, wild dreams, wild house cats, and wild literature. He associated it with other qualities: the good, the holy, the free, indeed, with life itself. By freedom he meant not rights and liberties but the autonomous and self-willed. By life he meant not mere existence but vitality and life-force.

Thoreau's famous saying, "in Wildness is the Preservation of the World," asserts that wildness preserves, not that we must preserve wildness. For Thoreau wildness was a given; his task was to touch it and express it, and he believed that myth expressed it best. Success was due not to political action or scientific study but to personal effort. *Walden* was his myth.

After Thoreau, our conservation ethic shifted from wildness to the preservation of wilderness, then biological communities, and, recently, to biodiversity. Wildness as a quality and its relation to other qualities is now rarely discussed.[2] This shift was broadly materialist—a move from quality to quantity, to acreage, species, and physical relations. The privileged status of classical science and its technologies in our culture virtually entailed this materialism. The world of classical science and its mathematics could not describe qualities like wildness, and what could not be described was ignored.

The shift was also reductive. By preserving things—acreage, species, and natural processes—we believed we could preserve a quality. But collections of acreage, species, and processes, however large or diverse, no more preserve wildness than large and diverse collections of sacred objects preserve the sacred. The wild and the sacred are not the kinds of things that can be collected. Historical forms of access and expression can be preserved, but one cannot put a quality in a museum. At the same time, wildness cannot disappear. It can be diminished in human experience, but it cannot cease to exist. The world contains many things that cannot be collected or preserved and put someplace—the set of complex numbers, gravity, dreams.

Apart from any relation to wildness, there are excellent reasons to preserve wilderness, biotic communities, and biodiversity, reasons that are thoroughly covered in our environmental literature. But the materialist and reductive shift in our conservation ethic has diminished the wildness of the places, species, and processes we have managed to preserve by diminishing their autonomy and vitality. And our conservation ethic tends to ignore this loss.

We continue to see the museum and its extensions—parks, wilderness, zoos, botanical gardens—as our models of preservation. In the past, political and aesthetic criteria selected the samples of acreage, the fauna and flora; in the future, one hopes, biological and ecological criteria will be foremost. But no matter how large the selection, the samples are rendered artificial by the process of selection. The environments (and their occupants) are managed according to human goals—the preservation of scenery, of resources, of wilderness, of biodiversity. Our artifice fundamentally alters their order by uprooting them from the context of interconnectedness that created that order. As Anthony Giddens says in discussing the consequences of modernity, "the end of nature means that the natural world has become in large part a created environment consisting of humanly structured systems whose motive power and dynamics derive from socially organized knowledge-claims rather than from influences exogenous to human activity."[3] This is just as true of parks and designer wilderness as it is of Disneyland.

Created environments have about them the aura of hyperreality so common in modern life. They are, says Jerry Mander, "all updated forms of Cain's desire to return home by remaking the original creation. The tragedy is that in attempting to recover paradise we accelerate the murder of nature."[4] Nature ends because it loses its own self-ordering structure, hence its autonomy, hence its wildness, hence its life. The stuff of museums exists, but it has no life.

Recently we have realized that our museums are too small and disconnected and artificial to preserve species and maintain their own structure and order. Our remedy for these island ecosystems and relic populations is to create even better created environments according to new knowledge about ecosystems and species. This leads to bigger and more complete ecosystems. It sustains some species, but it simultaneously diminishes their self-organization as the human influence and control mechanisms required for selection and preservation replace their own. The Wildlands Project is an example of this process and, if successful, could become the world's largest created environment. Its order and structure—the cores, corridors, buffers, and dense population areas—would undoubtedly be visible from space. I think of it as North America designed by Foreman, Noss & Associates.[5]

Many feel the pervasive Disneyesque and museumlike quality of wilderness areas, national parks, and wildlife preserves but continue to believe they provide a sanctuary from human artifice. This has always been an illusion. The national parks process millions of humans at the cost of natural processes. The wilderness of the Wilderness Act permits the state to control fire, insects, diseases, and animal populations, build trails for human use, graze livestock, and mine ore. These environments are not wild—they are too designed, administered, managed, and controlled.

Perhaps we need to imagine a new conservation ethic based on wildness. What we would come to mean by wildness could evolve from current interdisciplinary efforts by feminists, mathematicians, philosophers, and physicists to understand control, prediction, dominance, and their opposites—autonomy, self-organization, self-ordering, and autopoiesis.[6] It would provide a new understanding of wilderness in its original sense of self-willed land.[7] And it would promote Thoreau's project of understanding the wild within us and within nature as being fundamentally the same because of its association, conceptually, with vitality and freedom.

CONTROL VS. AUTONOMY

To construct a new conservation ethic we need first to understand why we impose a human order on nonhuman orders. The gain is prediction, efficiency, and control. Faced with the accelerating destruction of ecosystems and the extinction of species, we believe this is our only option. So we fight to preserve ecosystems and species and accept their diminished wildness. This wins the fight but loses the war. In the process we simply stop talking about wildness.

For instance, we substitute "wilderness" for "wildness." In Richard P. Primack's *Essentials of Conservation Biology*, Thoreau is misquoted as saying "in Wilderness is the Preservation of the World."[8] This substitution is so commonplace we don't notice it. But most of our designated wilderness is not in fact wild: the Gila, for instance, is a pasture, not self-willed land. Thoreu did not claim that in pasture is the preservation of the world.

We also equate wildness with biodiversity. In *Reclaiming the Last Wild Places*, by Roger L. DiSilvestro, the second chapter is entitled "Biodiversity: Saving Wildness," and there are phrases like "wildness in nature, . . . is what we preserve when we protect biodiversity," and "protection of biodiversity, of wildness."[9] But wildness is not biodiversity. Indeed, wildness may be inversely correlated with biodiversity. In the memorable case of the two oases in Gary Nabhan's *The Desert Smells Like Rain*, the oasis occupied by the Papago had twice as many bird species as the "wild" one preserved in Organ Pipe National Monument.[10] Neither is wild in any meaningful sense of the term. (Wilder desert oases might contain even fewer species.) But is wildness any less important than biodiversity?

For these writers the key distinction is between "in captivity" and "in the wild," which now means a managed ecosystem. But if grizzlies are controlled in zoos and controlled in wilderness, then what was for Thoreau the central question—wildness as freedom—simply drops out of the discourse of preservation. We also ignore wildness when we define wilderness in terms of human absence. In "Aldo Leopold's Metaphor," J. Baird Callicott says that, with the exception of Antarctica, there was no landmass without human presence, so the wilderness of the Wilderness Act is an "incoherent" idea.[11] If we fail to incorporate wildness into what we mean by wilderness we simply define wilderness out of existence.

Some people deny the existence of wildness on the grounds that any human influence on a species or an ecosystem destroys wildness. And since human influence has been around a long time . . . again, no wildness. One wonders what Lewis and Clark, standing on the banks of the Missouri, would have thought of such talk. "This isn't wilderness. Why, there are millions of humans out there. And it isn't wild either. Why, human influence has been mucking up this place for ten thousand years." Something is wrong here. Influence is not control and does not preclude autonomy. Autonomy is often confused with radical separation and complete independence. But the autonomy of systems—and, I would argue, human freedom—is strengthened by interconnectedness, influence, elaborate iteration, and feedback. Indeed, such self-organizing systems create that possibility of change without which there is no freedom. Determinism and autonomy are as inseparable as the multiple aspects of a gestalt drawing.[12]

The important point is that whatever kind of autonomy is in question—human freedom, self-willed land, self-ordering systems, self-organizing systems, autopoiesis—all are incompatible with external control, not external influence. To take wildness seriously is to take the issue of control seriously. And because the disciplines of applied biology do not take wildness seriously, they are littered with paradoxes: wildlife management, wilderness management, managing for change, managing natural systems, mimicking natural disturbance—what we might call the paradoxes of autonomy. Collections of paradoxes are usually bad news for paradigms.[13]

CONSERVATION BIOLOGY

Since Aldo Leopold, the biological sciences have played an increasingly imperial role in the conservation ethic. If that ethic is about preserving ecosystems and species, then one goes to the experts on ecosystems and species—ecologists and biologists. During the past twenty years it became obvious that the individual disciplines of applied biology were insufficiently comprehensive to achieve preservationist goals, especially biodiversity, and that they needed to be integrated with the newer disciplines of population biology and ecology—thus conservation biology. Conservation biology has become the dominant voice for preservation in this country, if not the world.[14]

Conservation biology is also about control. It is metamanagement. It integrates the controls available in the biological, physical, and social sciences. Since biodiversity is understood in the model as a scarce resource, the preservation of biodiversity becomes a problem in resource management.[15] In the face of biodiversity loss (and there surely is such a crisis), conservation biology demands we do something now in the only way that counts: more money, more research, more technology, more information, more acreage. Trust science, trust technology, trust experts; they know best.

This position mirrors the mode of response to crisis familiar from Michel Foucault's studies of insanity and crime. Like penology and criminology, conservation biology seeks control to pursue a mission: to end a crisis it has moralized. Although the maladies addressed by these disciplines have always been with us, and have been handled by other cultures in more imaginative ways, they are exacerbated by the conditions of modernity: overpopulation, urbanization, and pathological social structures. These disciplines strive to control symptoms, however, instead of striking at causes. Their controls are directed at the Other, not at social pathologies. Instead of remaking our societies they set about remaking the world and diminishing its autonomy.

The multiple meanings of *discipline* here are not accidental. The controls are always disciplinary in nature. They involve: capturing (shooting, darting, netting, trapping, apprehending, arresting); numerical identification (tattooing—from concentration camp inmates to grizzlies); technological representation (photography, X rays); chemical manipulation (of the brain in the mentally ill, of fertility in predators); surgery (lobotomies for the mad, and for predators the implantation of radioactive plaques to make their feces visible from satellites); monitoring (radio collars on animals, ankle monitors on prisoners); and constant surveillance to accumulate ever more information. Having inflicted all this upon the insane and the criminal, we now tend to inflict these manipulations on plants and animals. The forecast in Ecclesiastes—"For that which befalleth the sons of men befalleth beasts"— is confirmed.

Justified in the name of normality and equilibrium (or peace in our time), disciplinary technologies tend to develop into grand wars of salvation. Like the great wars for civilization, economics now wars against poverty, criminology now wars against crime and drugs. Despite pockets of success, the wars fail. Prisons create more criminals, the war on drugs creates more drugs, poverty and hunger have increased; but the failures neither discredit the disciplines nor halt their wars. Like Avis, they just try harder.

Conservation biology is in this tradition of grand salvation. It wants to conduct a war for biodiversity—hence its missions and strategies (from the Greek word for army—*stratos*) to remake the natural world according to its own vision. It will fail for the same reason the other disciplines fail: it does not strike at the causes of the malady but remains therapeutic. Its fondest hope is to arrest symptoms, and it presumes, desperately, that the malady is acute not chronic. Change comes from alteration of structure. The structure that a radical (*root*) position should focus on is the positive feedback system comprising overpopulation, urbanization, outrageously high standards of living, outrageously unjust distribution of basic goods, the conjunction of classical science, technology, the state, and market economics that supports the high standard of living, and the utter absence of a spiritual life. To *preserve* wildness, wilderness, and biodiversity is to attack that system of social pathology. And let's face it, most of us turn chicken in the face of such a challenge.

In ecology, the most powerful statement of a conservation ethic of controlling nature is Daniel B. Botkin's *Discordant Harmonies*.[16] Botkin presents graphic evidence of the devastation caused by unmanaged el-

ephants in Tsavo, one of Kenya's largest national parks. He argues tren-
chantly that our current ideas about nature are outmoded. He calls for
more management, more information, more monitoring, more research, more funding for education about the environment. He argues
for the preservation of wilderness primarily as a baseline for scientific
measurement. It is a powerful book. He concludes that "nature in the
twenty-first century will be a nature that we make; the question is the
degree to which this molding will be intentional or unintentional, desirable or undesirable."[17]

"A nature that we make." For most biologists and ecologists, the autonomy of nature is now a naive ideal and we must accept our new environmental maturity. The notions of climax, equilibrium, and stability we used to associate with ecology have been dropped, and with them
went any clear idea of what constitutes a healthy ecosystem. A recent
volume of essays on the subject suggests that "there exists considerable
basis for expanding consensus if the concept of health is given primary
identity as a policy concept."[18] This approach removes the health of nature as a property of the world, reduces it to human policy, and virtually
ensures that biologists and ecologists will go about fixing the world with
treatments and remedial actions. The Other is subsumed into current
social policy.

One would like to believe that the radical environmentalist can offer
something more, and different, from what mainstream environmental
thought can offer. This, however, is no longer obvious.

RADICAL ENVIRONMENTALISM

During the past five years, conservation biology has extended its influence to radical environmentalism by inverting themes that once legiti-
mized its radical content. Science, technology, and modernity were
once the problem. Now they are the solution, and I fear The Wildlands
Project may reduce *Wild Earth*, our best radical environmental publication, to the political arm of a scientific discipline.

Again the key issue is control and autonomy, not science and technology. Recent issues of *Wild Earth* and *Conservation Biology* have run
debates about the management of wilderness and wild systems, but
they haven't penetrated to the heart of the problem. Writing in *Wild*

Earth, Mike Seidman concludes his exchange by saying, "It seems that the depth of my critique of management went unnoticed." Seidman was being a gentleman.[19]

The autonomy of natural systems is the skeleton in the closet of our conservation ethic, although it is recognized that no one is dealing honestly with the issue. It appears in many forms. It explains the growing discontent with our control of predators: the elk hunts in Grand Teton National Park, the slaughter of elephants for management, the trapping and training of the last condors. It explains the increasing discontent surrounding the reintroduction of wolves to Yellowstone National Park. For a decade environmentalists fought for an experimental population; now, faced with the biological and political controls on experimental populations, many people prefer natural recovery.

Biological controls are ubiquitous. Biologists control grizzlies, they trap and radio-collar cranes, they have cute little radio backpacks for frogs, they even put radio transmitters on minnows. And always for the same reason: more information for a better ecosystem. Information and control are indivisible, a point made in great detail in James R. Benninger's *The Control Revolution*. It is the main point, perhaps the only point, of surveillance.

Perhaps we don't need more information; maybe the emphasis on biological inventories, information for designing wilderness recovery, surveillance, and monitoring is a step in the wrong direction. And what could possibly be radical about it? The Nature Conservancy has been doing it for years and the Department of Interior is going to do it too. Trying to be radical about public lands issues is like trying to be radical about laundromats. Indeed, it seems to me that radical environmentalism has backed away from many of its original stances. Perhaps the obsession with roads and dams betrays a crude, industrial idea of destroying nature and has blinded us to modern control technologies that imply even more potent modes of destruction. But instead of a critique of control we have deep ecologists like George Sessions and Arne Naess supporting, in principle or in practice, genetic engineering.[21]

Perhaps we just need big wilderness, not more technological information about big wilderness. Why not just set aside vast areas where we limit *all* human influence: no conservation strategies, no designer wilderness, no roads, no trails, no satellite surveillance, no overflights with helicopters, no radio collars, no measuring devices, no photographs, no GIS data, no databases stuffed with the location of every stone on the summit of Mount Moran, no guidebooks, no topographical maps. Let whatever habitat we can save go back to its own order as much as possible. Let wilderness again become a blank on the map.

MANAGING NATURE

There are two senses of preservation, and most preservationist efforts have followed the first—preservation as dividing, remaking, and moving nature according to our own desire for order. Doug Peacock presents the second sense with great clarity, calling biology "Biofuck" and saying, "Leave the fucking bears alone."[22] This echoes Abbey's "Let being be," which was a quote from Heidegger, who stole it from Lao Tzu, who put it nicely:

> Do you want to improve the world?
> I don't think it can be done.
>
> The world is sacred.
> It can't be improved.
> If you tamper with it, you'll ruin it.
> If you treat it like an object, you'll lose it.
>
> The Master sees things as they are,
> without trying to control them.
> She lets them go their own way. . . .[23]

Although most of the public believes this *is* the conservation ethic, leaving things alone is definitely the new minority tradition. But consider carefully the lines, "If you tamper with it, you'll ruin it. / If you treat it like an object, you'll lose it." What if the effect of scientific experts creating environments, treating ecosystems, and managing species is as bad, or worse, than the effects of unmanaged nature? In short, how well does managing nature actually work? Ecologists tend not to talk about this issue for fear of giving aid to the enemy.

In a recent essay entitled "Down from the Pedestal—A New Role for Experts," David Ehrenfeld presents several examples of predictive failure and the unfortunate consequences for natural systems.[24] Consider, for instance, the introduction of opossum shrimp into lakes in northwestern North America to increase the production of kokanee salmon: "The story is a complicated one, with nutrient loads, water levels, algae, various invertebrates, and lake trout all interacting. But the bottom line is that the kokanee salmon population went way down rather than way up, and this in turn affected populations of bald eagles, various species of gulls and ducks, coyotes, minks, river otters, grizzly bears, and human visitors to Glacier National Park."[25] Indeed, Ehrenfeld goes on to say that "biological complexity, with its myriad internal and external variables, with its open-endedness, pushes ecology and wildlife management a little closer to the economics . . . end of the range of expert reliability."[26] And this comes from the dean of conservation biologists. Do we really want to entrust the management of nature to experts whose reliability is akin to that of economists? This removes a bit of the glitter from the remaking and treating nature agenda. Ecologists are compared with economists because of their problems with prediction. Prediction, some think, is the essence of science: no prediction, no science; lousy prediction, lousy science. Unless, according to this view, the biological sciences can produce accurate, testable, quantitative predictions, they are well on their way to joining the dismal science. Well, if your idea of good science requires accurate prediction then all the sciences are looking a bit dismal, especially ecology.[27]

Historian of ecology Donald Worster, in an essay entitled "The Ecology of Order and Chaos," notes that "despite the obvious complexity of their subject matter, ecologists have been among the slowest to join the cross-disciplinary science of chaos."[28] This is not quite fair, but their lack of honesty on the subject probably has something to do with the unsettling consequences for the practical application of their discipline and hence their paychecks. They keep hanging onto the hope of better computer models and more information. As Bertolt Brecht said in another context, "If you're still smiling, you don't understand the news."

CHAOS AND COMPLEXITY

Most of the rapidly growing literature on chaos and complexity is either journalistic or extremely technical.[29] Of greater importance in this context are the philosophical implications of chaos and complexity. An excellent examination of these subjects can be found in Stephen H. Kellert's *In the Wake of Chaos*. Kellert suggests, as does Ehrenfeld, that the problems facing the practical applications of ecology and biology are more formidable than the disciplines are willing to admit.[30]

What emerges from recent work on chaos and complexity is the final dismemberment of the ideology of the world as machine. In its place is the view of the world characterized by wildness, vitality, and freedom, a view that goes well beyond Lao Tzu and Thoreau but one they would no doubt find inspiring. Most of nature turns out to be dynamic systems—not unlike a mean eddy line in Lava Falls, where the description of the turbulence (a nonlinear differential equation containing complex functions with *free* variables) prevents a closed-form solution. Such systems are unstable; they never settle into equilibrium. (Kayakers know this in their bodies.) They are aperiodic; like the weather, they never repeat themselves but forever generate new behavior, one of the most important of which is evolution. Life evolved at the edge of chaos, the area of maximum vitality and change.

Dynamic systems marked by chaos and complexity do have an order, and the order can be described with mathematics. They are deterministic and, usually, we can calculate probabilities and make qualitative predictions about how the system will behave in general. But with chaos and complexity, scientific knowledge is again limited in ways similar to the limits of incompleteness, uncertainty, and relativity. That does not end science. All that drops out is quantitative prediction, and that only affects most science in one way: control.

What happens to the rationality of managing species and ecosystems without accurate prediction and control? If the subsystems of an ecosystem (from vascular flows to genetic drift plus all the natural disturbances to ecosystems—weather, fire, wind, earthquakes, avalanches), if all these exhibit chaotic and complex behavior, and this behavior does not allow quantitative predictions, then isn't "ecosystems management" a bit of a sham and isn't the management of grizzlies and wolves at best a travesty? Why don't we just can the talk of health and integrity and admit, honestly, that it's just public policy? Why don't we just fire the Inter Agency Grizzly Team forthwith and let them do something useful?

Much of the best intellectual labor of this century has led to the admission of various limits of science and mathematics—the limits of axiom systems, observation, objectivity, measurement. This admission should have a humbling effect on all of us, and the limits of our knowledge should define the limits of our practice. The biological sciences should draw the line at wilderness—core wilderness, Wilderness Act wilderness, any wilderness—for the same reasons atomic scientists should accept limits on messing with the atom and geneticists should accept limits on messing with the structure of DNA. We are not that wise; nor can we be. The issue is neither the legitimacy of science in general, nor the legitimacy of a particular scientific discipline, but the appropriate limits on any discipline in light of limited knowledge. Accepting these limits and imagining a new conservation ethic based on wildness, vitality, and humble, careful practice would unite Thoreau's insight—"in Wildness is the Preservation of the World"—with ancient wisdom, the intuitions of our most radical wilderness lovers, the ecofeminists, and the cutting edge of mathematics and physics. The prospect is as consoling as it is charming.

Wildness is out there, the most vital hangs out at the edge of chaos, the wild Earth is radically free. Since God plays dice with everything, He must be a connoisseur of chaos. Perhaps we should join Him in this. The next time you howl in delight like a wolf, howl for unstable, aperiodic behavior in deterministic, nonlinear, dynamic systems. Lao Tzu, Thoreau, and Abbey—perhaps even God—will be pleased.

Notes

1. Robert D. Richardson, Jr., *Henry Thoreau: A Life of the Mind* (Berkeley: University of California Press, 1990), p. 225.
2. The notable exception is Gary Snyder's *Practice of the Wild* (San Francisco: North Point Press, 1990).
3. Anthony Giddens, *Modernity and Self-Identity* (Stanford: Stanford University Press, 1991), p. 144.

4. Jerry Mander, *In the Absence of the Sacred: The Failure of Technology and the Survival of the Indian Nations* (San Francisco: Sierra Club Books, 1991), p. 149.

5. Reed F. Noss, "The Wildlands Project Land Conservation Strategy," *Wild Earth* (Special Issue 1992):10–25.

6. Because of their interest in the issue of domination, much of the best work on control has been done by feminists. See Susan Griffin's *Woman and Nature: The Roaring Inside Her* (New York: HarperCollins, 1979), and Carolyn Merchant, *The Death of Nature* (San Francisco: Harper & Row, 1990). For the best discussion of autonomy, see Evelyn Fox Keller, *Reflections on Gender and Science* (New Haven: Yale University Press, 1985), pt. 2, chap. 5. On autopoiesis, see Humberto R. Maturana and Francisco J. Varela, *The Tree of Knowledge: The Biological Roots of Human Understanding* (Boston: Shambhala, 1992). On self-organizing systems see I. Prigogine and I. Stengers, *Order Out of Chaos* (New York: Bantam Books, 1984).

7. See the discussion of "wilderness" in Jay Hansford C. Vest, "Will of the Land," *Environmental Review* (Winter 1985):321–329.

8. Richard B. Primack, *Essentials of Conservation Biology* (Sunderland, Mass.: Sinauer, 1989), p. 13.

9. Roger L. DiSilvestro, *Reclaiming the Last Wild Places: A New Agenda for Biodiversity* (New York: Wiley, 1993), p. 25.

10. Gary Nabhan, *The Desert Smells Like Rain* (San Francisco: North Point Press, 1982), chap. 7. See also the introduction to Peter Sauer, ed., *Finding Home* (Boston: Beacon Press, 1992).

11. J. Baird Callicott, "Aldo Leopold's Metaphor," in Robert Costanza et al., eds., *Ecosystem Health* (Washington, D.C.: Island Press, 1992), p. 45.

12. John Biggs and F. David Peat, *Turbulent Mirror* (New York: Harper & Row, 1989), p. 74.

13. See Paul Hoyningen-Huene, *Reconstructing Scientific Revolutions: Thomas S. Kuhn's Philosophy of Science* (Chicago: University of Chicago Press, 1993).

14. Primack, *Essentials of Conservation Biology*, chap. 1.

15. See the diagram linking conservation biology and resource management in Primack, *Essentials of Conservation Biology*, p. 6.

16. Daniel B. Botkin, *Discordant Harmonies: A New Ecology for the Twenty-First Century* (New York: Oxford University Press, 1990).

17. Ibid., p. 193.

18. Robert Costanza et al., eds., *Ecosystem Health* (Washington, D.C.: Island Press, 1992), p. 14.

19. Mike Seidman's original letter appeared in *Wild Earth* 2(3) (Fall 1992):9–10. Responses to his letter by Reed F. Noss, W. S. Alverson, and D. M.

Waller appeared in *Wild Earth* 2(4) (Winter 1992/93):8–10. Seidman's reply is in *Wild Earth* 3(1) (Spring 1993):7–8.

20. James R. Benninger, *The Control Revolution: Technological and Economic Origins of the Information Society* (Cambridge: Harvard University Press, 1986).

21. Ariel Salleh, "Class, Race, and Gender Discourse in the Ecofeminism/Deep Ecology Debate," *Environmental Ethics* 15(3) (Fall 1993):233.

22. Rick Bass, "Grizzlies: Are They Out There?" *Audubon* 95 (September-October 1993):66–79.

23. Stephen Mitchell, trans., *Tao Te Ching* (New York: HarperCollins, 1988), chap. 29.

24. David Ehrenfeld, *Beginning Again: People and Nature in the New Millennium* (New York: Oxford University Press, 1993), pp. 148–50.

25. Ibid., p. 149.

26. Ibid.

27. For chaos and predictive failure in classical economics, see Richard H. Day, "The Emergence of Chaos from Classical Economic Growth," *Quarterly Journal of Economics* (May 1983):111–119.

28. Donald Worster, *The Wealth of Nature* (New York: Oxford University Press, 1993), p. 168.

29. The classic, of course, is James Gleick's *Chaos: Making a New Science* (New York: Penguin Books, 1987). See also M. Mitchell Waldrop, *Complexity: The Emerging Science at the Edge of Order and Chaos* (New York: Simon & Schuster, 1992). The most accessible introduction to the technical issues is Biggs and Peat, *Turbulent Mirror*. For discussions of chaos in fields ranging from ecology to quantum physics see Nina Hall, ed., *Exploring Chaos: A Guide to the New Science of Disorder* (New York: W. W. Norton, 1991).

30. Stephen H. Kellert, *In the Wake of Chaos: Unpredictable Order in Dynamical Systems* (Chicago: University of Chicago Press, 1993).

A Natural History of Silence

CHRISTOPHER MANES

Nature is a language—
can't you read?
—MORRISSEY

I thought I discovered silence in a grove of wild palms. Scattered here and there throughout the Mojave Desert, a few acres of fan palm oases—perhaps the rarest habitat on earth—endure with inexplicable patience the ten-thousand-year drought that made Southern California the tourist attraction it is today. Several of these oases lie just outside the city of Palm Springs, and in one of the many surreal juxtapositions of modern life, you can see the distant glimmer of Bob Hope's bronze-domed mansion while standing among centuries-old fan palms that may have descended directly from a seedpod dropped an ice age ago by a giant tree sloth slouching toward extinction. The fatuous and the unfathomable, within easy walking distance for a well-shod Cahuilla Indian.

These oases harbor what can only be called "desert silence." In the syllogism of the Mojave, sound requires motion, motion produces heat, and, therefore, desert creatures prefer to keep quiet. The dead calm forces you to hear the blood surge along your temples, a silence so pure it seems tangible, like a hand gripping your face.

Or so I thought. In fact, having been desensitized by the city's constant background noise, my ears simply weren't attuned to the many voices of a palm oasis: Jerusalem crickets wooing potential mates, hooded orioles admonishing intrusive neighbors, Gambel's quail warning their chicks of lurking bobcats. Nor could I possibly discern most of the lush undertones of meaning there, in the form of the inaudible shrieks of yellow-winged bats, or the chemical messages of carpenter bees, or the courtship rituals of giant palm-boring beetles. Far from being mute, a palm oasis, like all wild places, overflows with the conversations of creatures going about the business of living.

Like us, nature constantly emits signs. As David Abram puts it, we come to a world already full of "values, purposes, and meanings" belonging to entities other than humans.[1] Nonetheless, for centuries our culture has insisted that only humans are speaking subjects. The rest of the world exists only as inarticulate material, without a voice, without a soul, ego, mind, or whatever other philosophical term we use to describe the quality we associate with moral existence. *Dumban geschaft* is how medieval English clerics summed up this particular taxonomy: "dumb creation."[2]

But our world was not always dumbstruck. On the contrary, most societies throughout most of history considered nature replete with speaking subjects of the nonhuman kind, subjects that humans, for their own good, listened to and understood. Anthropologists call this cultural complex "animism," which, despite its mystical associations, at heart describes a linguistic and moral taxonomy different and more expansive than our own: people speak, and so do geckos, marmosets, yew trees, and waterfalls, each in their own way. Consequently, the way we now listen to or ignore nature is not some unchanging, universal given. Rather, nature's silence has a history, and in that history lie clues to address our culture's environmental dilemmas.

LISTENING TO NATURE

"People do not exploit a nature that speaks to them," said Hans Peter Duer.[3] The observation explains a great deal about the history of humanity's treatment of nature. At some point in the past, all cultures practiced animism, which means that all cultures experienced nature as a realm filled with speaking subjects requiring ethical treatment.

For animistic cultures, knowledge includes listening to nature in a quite literal way. Elk, thunderclouds, ravens, tumbleweeds, all produce signals intelligible to an animistic culture. A hunter who does not listen to his prey—in order to understand its likes, dislikes, needs, desires, and quirks—has a case of chronic malnutrition in his future. As John Dryzek puts it: "This necessity helps to explain the ecological sensibilities found in many preliterate societies: those without such orientations soon expire."[4]

Not all animistic cultures listened equally well to what their ecosystem expressed, and their lack of attentiveness, as the Easter Islands demonstrate, led to environmental abuse. Nonetheless, animistic cultures operate on the assumption that the nonhuman world contains independent, articulate, self-willed subjects who, like humans, communicate their purposes and values. Today we can scarcely imagine the vast dimension of connotations a Pleistocene hunter must have sensed overlooking a summer valley filled with the crescendos and diminuendos of Ice Age beasts. But his experience held the world and human cultures in good stead for uncounted millennia.

INTERPRETING NATURE

Sometime during the Middle Ages, however, nature ceased to be a subject and became—of all things—a symbol.[5] Specifically nature became a text. Thus, by the twelfth century, Hugh of St. Victor, the German philosopher, could talk about "the Book of Nature" (*lex naturalis*) and be understood by his audience, as he probably would not have been by his animistic Teutonic ancestors, or by citizens of classical Greece, or even by eleventh-century Europeans outside a monastery.[6]

This powerful metaphor captured Europe's understanding of nature. It prompted Shakespeare to tell an audience made up, in part, of illiterate laborers that they could find "books in the running brooks / Sermons in stones."[7] It convinced Sir Thomas Browne to seek an understanding of God not only in the Bible but in nature, "that universall and public Manuscript, that lies expans'd unto the eyes of all. . . . Surely the Heathen knew better how to joyne and read these mysticall letters, than wee Christians, who cast a more careless eye on these common Hieroglyphicks, and disdain to such Divinity from the flowers of nature."[8]

These "common hieroglyphs" included the complete animal and plant kingdoms, which, like paintings on Egyptian tombs, had to be deciphered for their true, otherworldly meanings. An entire literary genre, the bestiary, thrived during the Middle Ages to provide these animal "interpretations." In the Anglo-Saxon poem "The Whale," for example, the poet insists that the cetacean pretends to be an island in order to trick unwary mariners onto its back, whereupon it drags them to their doom at the sea bottom.[9] The meaning: whales represent the biblical leviathan, which in turn is a figure of Satan, who beguiles men's souls into a false sense of security that plunges them down to hell.

In this taxonomy of nature, humans were independent speaking subjects, but nature constituted a series of mute symbols whose grammar expressed divine meanings, just as words in a book reflect an author's intent. As a result, the medieval person did not listen to nature, he *interpreted* it, the same way a reader must interpret this sentence to grasp my meaning. For this reason, it never occurred to the author of "The Whale" to actually observe a whale to find out if its behavior conformed to bestiary legend: since the animal had no volition of its own, whales had no choice but to beguile sailors in order to make a divine point. To admit otherwise would have been tantamount to an assertion that the words on this page have the power to rebel against the author and speak their own independent messages.

Thus a twelfth-century French monk, looking out at the same warm summer landscape as his Ice Age ancestor, did not perceive the innumerable purposes of the flora and fauna, but rather only a complex grammar expressing a single purpose, that of the world's author: God.

INTERROGATING NATURE

The experimental method developed by Galileo in the sixteenth century, and the scientific revolution of the seventeenth century, closed the book on the Book of Nature. Granted, even today we understand the textual metaphor, but we experience it as a metaphor, not as the reality medieval institutions insisted it must be.

The institutions that came to dominate the Renaissance, particularly the mercantile institutions, understood that their power derived not from interpreting a moral universe but from manipulating the physical world. The whaling industry, for example, unlike the clerical author of "The Whale," had little interest in a moral discourse about cetaceans representing Satan. Unlike the church, the whaling industry depended not on tithes, aristocratic bequests, and the grandeur of Scholastic apologia, but rather on whale oil. Therefore, the industry required particular knowledge about the real, veritable habits of whales in order to maximize whale oil yields and hence augment the industry's profits and influence. "*Nam et ipsa scientia potestas est*," said Sir Francis Bacon: knowledge is power.[10] Science provided a new discourse for mercantile power, and a new regime of silence for nature took hold.

The philosopher of science Hans Reichenbach writes that "an experiment is a question addressed to nature; by the use of suitable devices the scientist initiates a physical occurrence the outcome of which supplies the answer 'yes' or 'no' to the question."[11] The experimental method thus introduced a new limit to the speaking of nature: the limits of an interrogation. For the scientist-inquisitor, nature has no voice of its own; it can merely respond to human interrogatories with a simple yea or nay. The rest of the buzzing, howling, snorting, clamor of nature constitutes mere babble.

The interrogation of nature has continued for many centuries now, compiling vast libraries of binary knowledge but very little understanding of nature. Primal peoples discovered millennia ago that clearcutting forests damages the complex relationships that maintain biological diversity and hence healthy human communities; modern science, in its interrogative mode, only recently put two and two together on this issue. Both primal hunters and modern scientists employed "reason" to come to their conclusions, but the former listened to nature and its purposes, while the latter merely demanded answers. As a result, the people who formulated the questions (in this example, the forest industry and related government agencies) determined in advance the type of knowledge worth learning, which excluded the simple but necessary: "What do you, the forest, want?" That question required more than a yes or no.

POSTMODERN NATURE

Today, inquisitorial science faces the challenge of its successes. While as a practical matter the scientific method has remained unchanged since Galileo's time, its intellectual underpinnings have given way under the scrutiny of postmodernism. This has occurred mostly from within the ranks of scientists, who have grown suspicious of the power so casually wielded by science. Thus, for instance, Heisenberg's principle of indeterminacy has made obsolete the notion of a detached scientist posing questions through experimentation to a passive natural world. In fact, every competent scientist now knows that observation, measurement, or recordation of nature alters the thing observed. This applies not only on the nano-level of particle physics, but on the macro-level of rhinoceroses and redwoods. An experiment on an animal in a laboratory has already changed the character of the animal experimented upon, as well as the scientist performing the experiment.

Consequently, some scientists have begun to recognize that science is less like an interrogation and more like a conversation between two articulate subjects. Biologist Charles Birch, for example, envisions a "postmodern biology" which extends subjectivity to nonhuman entities.[12] Similarly, Evelyn Fox Keller, when writing about the eminent geneticist Barbara McClintock, observes that "the objects of her study have become subjects in their own right; they claim from her a kind of attention that most of us experience only in relation to other persons."[13] Fifty years ago, these statements would have been not only controversial but unintelligible, just as Hugh of St. Victor's metaphor, lex naturalis, would have perplexed an eleventh-century audience.

Moreover, the science of ecology itself has undermined inquisitorial science. From its putative origins in the natural history studies of Thoreau, research into the interrelationships among biological communities presaged the need for a broader vision of science. The desire to communicate with, rather than cross-examine, nature continued with Muir, Leopold, and Carson. Most recently, conservation biology has vigorously rejected the fiction of indifferent scientists with no connections to the natural communities they study beyond the reach of a syringe. As David Abram suggests in his writings, we may be seeing the development of ecology as the matrix for all other sciences, a matrix in which nature matters again to a scientific community that listens to a world filled with communicative subjects.[14]

Paradoxically, we begin to come full circle: from animism which heard voices in the wilderness to postmodern biology which seeks communication with the subjects of nature. This postmodern science promises to return us to our senses—to a recognition of the meanings and purposes of the cricket, oriole, quail, bat, and beetle of the palm oasis. After wandering in a desert of moral silence about nature for centuries, our culture now must have the good sense to enter the oasis, sit still for a moment, and listen.

Notes

1. David Abram, "The Perceptual Implications of Gaia," *The Ecologist* 15 (1985):96.

2. George Krapp and Dobbie Van Kirk, ed., *The Anglo-Saxon Poetic Records* (New York: Columbia University Press, 1979), vol. 3, p. 1127b.

3. Hans Peter Duer, *Dreamtime: Concerning the Boundary Between Wilderness and Civilization*, trans. Felicitas Goodman (Oxford: Basil Blackwell, 1985), p. 90.

4. John S. Dryzek, "Green Reason: Communicative Ethics for the Biosphere," *Environmental Ethics* 12 (1990):210.

5. This is, of course, a simplification and a generalization. Like all other forms of knowledge in our culture, views of nature take complex and often contradictory forms, sometimes agreeing with, sometimes opposing, institutions with a stake in one view or another. Nonetheless, the medieval church had more power over written discourse, and hence future culture, than, say, a medieval farmer did. See Michel Foucault, *Madness and Civilization*, trans. Richard Howard (New York: Random House, 1973), for a discussion of the relation between institutional power and shifts in knowledge.

6. Hugh of St. Victor, *The Didascalicon of Hugh of St. Victor: A Medieval Guide to the Arts*, trans. Jerome Taylor (New York: Columbia University Press, 1961).

7. *As You Like It*, act II, scene i, lines 16–17.

8. Sir Thomas Browne, *Religio Medici and Other Works*, ed. L. C. Martin (Oxford: Clarendon Press, 1964), p. 15.

9. "The Whale," in Krapp and Dobbie Van Kirk, *The Anglo-Saxon Poetic Records*.

10. *Meditations Sacrae*, 1597.

11. Hans Reichenbach, *The Rise of Scientific Philosophy* (Berkeley: University of California Press, 1968), p. 97.

12. Charles Birch, "The Postmodern Challenge to Biology," in David Ray

Griffin, ed., *The Reenchantment of Science: Postmodern Perspectives* (Albany: State University of New York Press, 1988), pp. 69–78.

13. Evelyn Fox Keller, *A Feeling for the Organism: The Life and Works of Barbara McClintock* (San Francisco: W. H. Freeman, 1983), p. 200.

14. Abram, "Perceptual Implications."

The Rhizome Connection

DOLORES LACHAPELLE

Fundamental to the Plains Indians' concept of their world is the teaching of the Six Powers: the Powers above, the Powers below, and the Powers of the four directions. The tribe lived within these sacred powers. Our Eurocentric culture trivialized the Powers of the four directions into mere points on the compass, concentrated the Powers above into one single male god up there in the sky, and turned the Powers below into Hell—where the damned go. With innovative research on the forest as an interconnected system—not just a collection of trees—modern peoples are finally beginning to recognize the "Powers below." We are beginning to see that they may well be the most important of all, not only for human life but for all life on earth.

The metaphor of the tree has "dominated Western reality, and all of Western thought, from botany to biology and anatomy, and also gnosticism, theology, ontology, philosophy," according to the French philosophers Deleuze and Guattari. The tree leads to a hierarchical order of a central trunk with larger branches and smaller branches. The trunk forms the connection between all the parts, thus in a way limiting connections. The rhizome is quite different. Any point on a rhizome can be connected with any other. A tree can be cut down, but rhizomes are much less subject to destruction. A rhizome can grow again along another line if broken at any point.

A rhizome is a thickened stem for food storage that grows horizontally along or under the surface of the soil. Roots, flower stalks, and foliage grow from buds on the rhizome. The rhizome spreads underground and comes up wherever it wants, often providing humans food

An earlier version of this essay appeared in *Wild Earth* (Fall 1993).

without plowing up the land: a true giveaway from nature to us. Rhizomes produce food for much of the world outside European areas. Taro is the main food plant for Polynesia and Hawaii, and it grows also in Japan and China. In fact, rice began as a weed in taro patches. Banana trees are a rhizomatous plant now growing in Africa, Asia, Australia, and Latin America. Bamboo, the main structural material for many Asian cultures, is another rhizomatous plant. Manioc (cassava) is a prime food in South America and cattails feed Indian cultures in North America.

Perhaps the very structure of a healthy relationship with nature depends on the free and easy giveaway by rhizome plants of food to humans, as opposed to the hard work and uncertainty of agriculture. If you can count on a taro plant to send up new plants continually, your relationship to nature is one of eternal gratitude and trust. With agriculture came doubt, worry, and prayers. Consider crabgrass, a rhizomatous plant related to millet. It is an extremely successful weed, as any suburban, perfectionist lawnperson knows. You can't get rid of it. Compare that to a food plant we must continue to care for and worry over.

Deleuze and Guattari have developed an entire system of thinking based on the rhizome. They ask: "Isn't there in the East, noticeably in Oceania, a kind of rhizomatic model that contrasts in every respect with the Western model of the tree?" They wonder if this is "reason for the opposition between the morals of philosophies of transcendence dear to the West and those of immanence in the East" and continue: "The demands made on the mind are, like this God's name, unspeakable. Brain and conscience are commanded to vest belief, obedience, love in an abstraction purer, more inaccesible to ordinary sense, than is the highest of mathematics." Trying to please this "unbearable absence" leads to an unending search for perfection, which in turn has led to the unending destruction of nature. In cultures based on the rhizome as food, including most so-called primitive cultures, the gods are multiple and ever-present. The gods are like the rhizome, "which connects any point with any other point." The Tsembaga of New Guinea have a whole class of gods based on the word *mai*, which is their word for the rhizome of the taro plant. These gods or spirits of the low ground are involved in the cycle of fertility, growth, and decay. But decay is not considered merely death, as we think, but rather the source of more life.

Deleuze and Guattari give this advice: "Form rhizomes and not roots. Be neither a One nor a Many, but multiplicities! A rhizome doesn't begin and doesn't end, but is always in the middle, between things, interbeing. The tree is filiation, but the rhizome is alliance, exclusively alliance. The tree imposes the verb 'to be' but the rhizome is woven together with conjunctions: *and. . . and. . . and.*" Deleuze and Guattari explain the advantage of this kind of metaphorical structure for revolution, but I say it's even better for devolution. All beings, human and nonhuman, are connected in the bioregion through the Powers below.

Deleuze points out that in psychoanalysis the usual tree metaphor leads to "central organs, the phallus, the phallus-tree." This limits psychoanalysis because there is always a "general or a boss in psychoanalysis (General Freud)." Cut off from the rhizome of our place—the interlocking connections we have with rocks, plants, and animals of that place—our culture's sexual traumas and phobias are almost inevitable. In a truly place-centered culture, a vast network of interlocking rhizomes links each to all.

Jung began the breakout of the narrowly human, Freudian self when, near the end of his life, he wrote: "Life has always seemed to me like a plant that lives on its rhizome. Its true life is invisible, hidden in the rhizome. The part that appears above ground lasts only a single summer. . . . What we see is the blossom, which passes. The rhizome remains."

A further development occurred when Arne Naess defined what he calls the ecological Self: "We don't like the distinction between humans and environment. . . . We take the ecological view where you are in a network, in which you cannot single out anything, an interrelated network which is intrinsic." A true rhizome network!

Recently, the radical Jungian James Hillman wrote: "Maybe the idea of self has to be redefined." He points out that the usual self, as defined in psychology all these years, "is the interiorization of the invisible God beyond. . . . I would rather define self as the interiorization of community . . . something more ecological . . . a psychic field. And if

I'm not in a psychic field with others—people, animals, trees—I am not." Hillman continues: "Therapy—even the best deep therapy—contributes to the world's destruction. We have to have new thinking—go back before Romanticism, and especially out of Western history, to tribal animistic psychologies that are mainly concerned, not with individualities, but with the soul of things ('environmental concerns,' 'deep ecology,' as it's now called) and propitiatory acts that keep the world on its course." The rhizome represents deep connections through the very ground of the place.

How do we get back down to our grounded rhizome connections? I'll give you two examples—one, how I personally discovered rhizomes; the other, an indigenous myth. I first noticed the power of rhizomes after moving to Silverton, Colorado, twenty years ago. The very soil of the town site has been trashed by old smelting dumps and modern big machines; yet when the first miners came here they called it Baker's Park, it was so beautiful. Amidst the smelter tailings and scraped soil, I kept seeing patches of the lovely wild Rocky Mountain iris. When I dug iris up to transplant, I discovered it was a rhizome—which keeps on growing and spreading and blossoming in the refuse of this old high-altitude mining town.

"Swamp dancers," cattails, show us the way to the rhizome connection through myth. This rhizomatous plant can provide us with many of life's necessities, as it did for Native American tribes in most parts of the continent. Each part of the plant is edible at some time from early spring until snow flies. This is how Doug Elliott retells the myth for us:

According to an old Indian story, Coyote—that foolish trickster—was out walking one evening. He was following a trail along a ridge beside a low-lying area when he suddenly heard sounds coming from that low spot. It sounded like music. It was almost dark and he couldn't see very well, but he could hear the music. Yes, there was no mistaking; that was the sound of dance rattles. There was a dance going on.

There was nothing Coyote liked better than a dance. He knew he was the best dancer and he loved to get right in the middle of the dance circle and show off, so he hustled right down that hill and pushed his way into the middle of the crowd. He could hear the rhythmic rustling of the dance rattles. Swish, swish, swish, swish. There were so many of them and they were making beautiful music. Swish, swish, swish, swish. Everybody was swaying back and forth. Coyote started doing his fanciest steps and saying things like, "You think you know how to dance. I'll show you how I dance!" He started really strutting his stuff, but nobody seemed to notice. They just crowded around him and kept swaying back and forth. And the dance rattles kept their steady rhythm.

Coyote was getting tired, but he didn't want to be the first to quit. Finally he said, "You know, sometimes when we dance, we rest!" But nobody stopped to rest, and when the dawn came, he realized that he had been dancing with swamp dancers.

That's the Native American name for cattails. And as Coyote looked around, bleary-eyed and exhausted, the swamp dancers were still swaying back and forth. They're still out there and they are still dancing.

References

Deleuze, Gilles, and Felix Guattari. *On the Line*. New York: Columbia University Press, 1983.

Elliott, Doug. "Swamp Dancers." *Wildlife in North Carolina* (Spring 1993): 17–19.

Hillman, James, and Michael Ventura. *We've Had a Hundred Years of Psychotherapy and the World's Getting Worse*. San Francisco: HarperCollins, 1992.

Jung. C. G. *Memories, Dreams, Reflections*. New York: Vintage Books, 1961.

Steiner, George. *In Bluebeard's Castle: Some Notes Towards the Redefinition of Culture*. New Haven: Yale University Press, 1971.

A New Vision for the West

GEORGE WUERTHNER

One of the dilemmas we face in the West is the obvious failure of the present reserves system to protect wildlife species, biodiversity, and ecological processes. It is estimated, for example, that two thousand grizzly bears is the minimum viable population necessary to maintain grizzlies in the northern Rockies for the foreseeable future. Greater Yellowstone at 18 million acres, one of the largest remaining temperate wildlife complexes in the United States, supports two hundred and fifty or fewer bears today.

Although there is plenty of room for expansion of bears into presently unoccupied regions like the Salt River and Wyoming ranges and the numerous aspen-dotted ranges of the Caribou National Forest, it is unlikely that even at maximum habitat utilization Greater Yellowstone alone could sustain two thousand bears. If bears are to survive here, we must expand their range beyond the public and private land sectors of Greater Yellowstone, including northward to other occupied bear habitat in the Glacier–Bob Marshall Ecosystem and southward into the Unitas on the Utah–Colorado border. An estimated minimum of 12 million acres is necessary in Greater Yellowstone to provide a sufficiently large region to support ecologically viable fire regimes. At 18 million acres, Greater Yellowstone is barely large enough to provide for such landscape-wide events.

These insights have convinced many conservation biologists and some "conservationists" that the majority of our reserves, including most of our national parks, wilderness areas, and even much smaller natural areas like Nature Conservancy reserves, are of insufficient size to maintain viable evolutionary functions, ecological processes like wildfire, and populations of wildlands-dependent species like the grizzly. These failures are often cited by critics of conservation efforts as examples of "how you can't put a fence around it and expect to protect wildland processes and landscapes." While not denying that human influence is everywhere—that smog from California's Central Valley threatens the Sierra and acid rain carried from the Midwest has sterilized lakes in the Adirondacks—the fact that our present preserve system does not work as well as it should does not mean it could not work. While the two hundred and fifty grizzlies left in Greater Yellowstone may not be sufficient to maintain long-term population viability, it must be noted that at least there are grizzlies left in Greater Yellowstone—suggesting that wildland preservation, on a sufficiently large scale, can maintain wildland ecosystems and native species.

Nevertheless, there is no denying that our current array of wildland reserves, with few exceptions, are simply too small to provide for the ecological needs of wide-ranging species and the maintenance of ecological processes. They are like airplanes with wings and engines too small to become airborne. If our present system of reserves fails, it is not, I would argue, that we don't have a good idea of how to make them fly.

USE OF THE LAND

The current problem stems from the fact that the majority of the public lands, and nearly all of the private lands, are devoted to commodity exploitation and production. The lands devoted exclusively to ecosystem protection and function make up a relatively minor portion of the landscape. Is the solution to make all resource extraction kindler and gentler so that it is more compatible with wildland ecosystem preservation? Or should we consider expanding the lands devoted to wildland preservation and shrink the amount of the landscape devoted to consumptive resource utilization? I believe the latter option has the best potential for long-term ecosystem preservation and integrity. We need to think big and design reserves that are continental in scale and scope. To accomplish this requires changing attitudes as well as the way we use the landscape. However difficult this may be, I believe it is much more likely to be successful than any schemes we have yet developed to make landscape exploitation compatible with wildland ecosystem preservation.

Those opposed to such an expansion suggest that the vision is impossible for a number of reasons. They argue that humans have always been a part of natural ecosystems and thus should not be prevented from exploitation now. Furthermore, it is argued, humans already use much of the earth, and unless we can integrate humans into the landscape, we will never protect anything. We need "wise use," not "no use."

The people-are-a-part-of-nature proponents argue that human exploitation is a "natural" process. By such a definition, cutting down the forests of the world is natural because humans have always cut down trees and forest ecosystems have persisted. Similar arguments are made for domestic livestock grazing, hunting, and a host of other consumptive resource uses. This argument conveniently overlooks the size, scale, and scope of current human exploitation, which is on a magnitude thousands of times greater and more intensive than in the past. Although I believe pre-Columbian people probably had far more impact on the landscape than most people give them credit for, their limited technology, combined with low populations, restricted the effects of their exploitation to a level that did not, in most instances, jeopardize individual species, much less entire ecosystems.

A less extreme, more subtle, and hence more pernicious argument, now frequently advocated by even the conservation community, is "sustainable economics." The idea behind sustainable economics is that we live off the "interest" not the "capital" of the landscape. If exploitation of ecosystems is pursued in moderation, humans should be able to skim off some amount of any resource—timber, fish, wildlife, grass—without damaging the ecological processes that produce them. If done properly, therefore, we theoretically should be able to cut a few trees indefinitely from the forest without jeopardizing the forest ecosystem. Such an approach is of course necessary if the human species is to survive at all, but we still don't know how to create sustainable extractive industries. And even if we did, a prudent approach would dictate that we should not apply sustainable development (or any other

kind of development) to every corner of the globe. In far too many instances, sustainable development is used to justify the status quo. So we
are told that sustainable development will preserve farming, logging,
fishing, or ranching communities intact, whether they deserve to be
maintained or not.

"NEW" MANAGEMENT

New Forestry, for instance, is hailed as one technique designed to provide timber to keep mills operating while still providing sustainable
forests. The idea is to remove only some trees while leaving more snags,
down timber, and woody debris than traditional forestry methods have
done in the past. But we don't really know if it's possible to extract trees
from the forest in any kind of economically viable fashion and still
maintain forest ecosystems. Although we may someday be able to produce enough trees to sustain a local mill indefinitely, this is not the same
as preserving a forest ecosystem.

We have not had enough years to evaluate whether New Forestry is
any better than old-fashioned forestry in the long run. We would be
wise to remain cautious in our applications of New Forestry or any
other kinder and gentler method of exploitation to large portions of the
landscape. After all, every new resource extraction method has been
hailed as a new breakthrough that will ensure long-term sustainability
of the exploited resource. A generation of foresters were taught—and
truly believed as strongly as today's advocates of New Forestry—that
clearcut logging would provide sustainable forests. No one advocates
the destruction of forest ecosystems. Yet it took thirty to forty years and
the near extinction of wild forest ecosystems before the ecological failures of this method became apparent. What will we learn about New
Forestry's failures in future years?

Similarly, advocates of new rangeland methods like Allan Savory's
holistic resource management (HRM), which purports to mimic the
migratory effects of bison by moving domestic livestock rapidly from
pasture to pasture, have a similarly myopic view of rangelands. We
barely understand how livestock grazing affects soil organisms, lichen
crusts, invertebrates, and nutrient cycling. To presume that mimicking the movement of one animal results in preserving the integrity of an

entire ecosystem represents shallow ecological thinking. Proponents of HRM frequently justify livestock grazing by asserting that grassland ecosystems need to be cropped to be "healthy." Despite the happy coincidence that HRM conveniently justifies continued livestock production, it ignores the fact that there are, and have always been, grasslands (on buttes, steep hillsides, inside canyons, and elsewhere) that have never been grazed by large ungulates and yet remain healthy, suggesting that health is somewhat subjective. Even if cropping were necessary, our native species do cropping more efficiently than domestic livestock. With native species, grassland ecosystems are divided up among species so that antelope, elk, bighorn, bison, prairie dog, and the rest all utilize different parts of it. The end result is far higher biomass in native animals than is ever achieved using one kind of domestic animal—another point conveniently ignored by HRM proponents.

LIMITED KNOWLEDGE

The point of all this is that our understanding of ecological processes is not only extremely limited but biased by the needs of commercial extractive industries. You don't get much money to research how forestry-disrupted ecosystems function, to discover how modern recreational hunting negatively affects wildlife, or to investigate the ecosystem degradation that results from livestock production. Because of these limitations, we can't fairly assess the impacts of these activities upon ecosystems. Even if this knowledge were available, there is no evidence that we would use it rationally. Is it rational to grow a water-dependent animal like the domestic cow in the arid West?

Should we not find some way to exploit the landscape in a more benign manner? All species exploit the environment to a degree. If we humans are to live, we must consume resources—but it is a matter of to what degree, where, and how. Let us not deceive ourselves. Having a sustainable supply of timber, farm products, or beef—the goal of resource exploitation industries and communities dependent upon them—is not the same as maintaining fully functioning wild ecological communities. Prescribed burns do not substitute for the variability provided by wildfires. Cows do not emulate bison. Hunters do not substitute for native predators. It is doubtful we can mimic the infinite variations found in wildland systems. There is no substitute for wild eco-systems and native wildlife and flora. Kinder and gentler resource exploitation has its place, but the bottom line is this: there is no human-created replacement for wildland ecosystems.

SOLUTIONS

So what's the solution? How can we protect wildland ecosystems and expand them if present reserves are too small to function properly? How can we meet current human demands for natural resources? Obviously we need to grow some crops if we are to feed ourselves, cut some timber if we are to continue building wooden houses, and raise cattle if we continue to consume beef. But do we need to raise beef cows in Nevada or even consume beef in the first place? Is it necessary to plow up the short-grass prairie in Montana to raise wheat or grow corn in Iowa to feed cattle and pigs? Should we log commercial timber on the steep slopes of the Idaho Rockies? Such questions are seldom discussed among policymakers or even many environmental organizations. The underlying assumption is that the landscape we now manipulate must, and will, continue to be used for resource commodity production.

Yet, at a minimum, it can be argued that our public landholdings in the West might serve more purposes if we devoted them to nonextractive resource functions. Our public lands are the last place where we can reasonably expect ecosystem protection and function to have priority across large landscapes. Isn't this, after all, the best use of these lands from a national perspective? Yet we squander this potential by permitting public lands to be dominated by marginal extractive industries. For instance, some 250 million acres of western public lands—an area equal to all the eastern seaboard states combined—are committed to livestock production, yet these lands provide only 3 percent of the forage used to produce the nation's beef. Nevada is an extreme example of this limited production. Nearly all of its public lands (50 million acres—bigger than all of New England) are devoted to livestock production, yet the state only produces as much beef as tiny Vermont. Is this a good use of land? Does it make sense to compromise vast areas of Nevada's natural landscapes and destroy much of the state's native bio-diversity to produce so little? When we consider that much of the agricultural cropland in the West as well as the Midwest is committed to grain production ultimately fed to domestic livestock, it becomes questionable whether we should be growing beef anyplace at all. And certainly we can question whether this should be part of any sustainable ecosystem or economic strategy.

Similar limitations exist for timber production and crop production in much of the West as well. The entire timber production in the northern Rockies on both public and private lands, for instance, contributes only 1 percent of the annual timber cut nationally. And even at this rate of harvest, we are cutting trees faster than the rate of replacement. Yet to produce this paltry nonsustainable 1 percent of the nation's wood supply, we are roading and carving up a significant portion of the last large wildland ecosystems in the northern Rockies.

Certainly it is worth questioning whether we should be attempting to sustain western communities by promoting marginal livestock, farming or logging, and nonsustainable industries. All these activities are consumptive of wild landscapes. Many of them are marginal in product output and contribute little to the overall national supply consumed by people. Agricultural production (including livestock) is easily the biggest threat to the ecological integrity of much of the western United States, yet it receives scant attention from environmental groups. Indeed, a review of the species extirpated or severely reduced in numbers in Montana alone—grizzly bear, wolf, bighorn sheep, prairie dog, black-footed ferret, swift fox, Columbian sharptail grouse, sage grouse, Yellowstone cutthroat trout, Westslope cutthroat trout, arctic grayling, and a host of others—reveals that all have suffered declines primarily due to agricultural impacts.

If we wish to restore wildland ecosystems—at least in the West—some reduction in agricultural production poses the greatest opportunity for significant land restoration. Since much of this agricultural production is sustained by government subsidies including crop supports, drought insurance, emergency livestock feed, subsidized irrigation, subsidized public lands grazing, taxpayer-funded predator control, and a host of other programs, it also suggests that much of this ecosystem degradation resulting from western agricultural practices is unsustainable economically. This is a reflection of landscape limitations prevalent over much of the West, including aridity, short growing season, frequent drought, serious soil erosion, rugged terrain, and other factors that make much of the West unsuitable for agricultural production—without subsidies. In 1992, Montana farmers received more than $320 million in direct government subsidies, accounting for almost 17 percent of their gross income. And this was in a good year when government payments were significantly less than in years with crop failures. Other subsidies are more subtle but perhaps even more significant than actual cash payments. Nearly 60 percent of the agricultural lands in Montana are considered highly erodible. If the costs of soil erosion were included in the true cost of wheat and barley production in Montana, most Montana grains would be too expensive to grow.

Growing cattle in the western United States requires huge environmental subsidies as well. The natural aridness of the West directly affects livestock production in many ways. Cattle, a water-dependent species, spend great amounts of time in riparian areas—those thin green lines of vegetation found along streams and lakeshores. Riparian areas are biologically critical for up to 80 percent of native western species, but they make up only 1 percent of the western landscape. Yet livestock, by concentrating their feeding in these areas, not only consume much of the vegetation but trample the banks and compact the soils with disastrous results for native species and physical processes, such as flood control, provided by healthy riparian areas.

Riparian areas may also be destroyed even if livestock never set foot in them. To provide sufficient forage to overwinter lifestock usually requires irrigated hay production. Irrigated forage for livestock is the major consumer of water throughout the West, even in heavily urbanized California, where the water used to provide irrigated pasture and alfalfa would supply the needs of 46 million urban dwellers. Water pumped from streams or underground aquifers is reducing the zone of riparian vegetation and destroying aquatic habitat for fisheries and aquatic invertebrates. Irrigation is a major factor contributing to the decline of native fish and the reason why more than four-fifths of all native western fish species are currently endangered.

These are some of the costs of livestock production that are not actualized in the cost of production. All in all, you can raise livestock with fewer environmental costs in the East than in the West. But presently most western livestock producers do not pay for these environmental costs, so they can compete with eastern producers. The problem with such subsidies is that they promote environmentally destructive land uses, discourage innovation, and never end. Since much of Montana as well as the rest of the arid West is fundamentally unsuitable for wide-spread agriculture, whether farming or livestock production, the only way farmers and ranchers can continue to survive is with continued government subsidies.

SUSTAINABLE COMMUNITIES: THE ALASKAN MODEL

What makes human communities sustainable? Perhaps we need to approach the idea of sustainability differently. When we look at natural systems, we find they fluctuate all the time. If there is a "balance" out there, it is found in amplitudes between means and is continuously changing. Due to climate and other factors, ecosystems alter in time and rarely remain static. Perhaps human communities operate in the same fashion.

We could adopt what I call an "Alaskan landscape" model as opposed to the "midwestern agricultural" model. In Alaska you have a few large urban centers, like Anchorage and Fairbanks, and some small villages and towns scattered throughout the rest of the state. But the overwhelming feature of Alaska is still wilderness. Instead of wildlands surrounded and fragmented by civilization and developed landscapes, you have a landscape where wilderness surrounds the civilization. Such a model is possible to recreate in much of the West, as well as other parts of the country, such as the northern forests of New England and the Upper Midwest. All that is required is reduction or elimination of land-consuming industries such as farming, livestock production, or commercial logging.

This approach would have three positive results. Look at what might happen to timber production if marginal lands were removed from the timber base. First the amount of land devoted to marginal timber production, which is usually the most environmentally sensitive, would be reduced. This would permit expansion of wildland landscapes. It would also reduce the amount of marginal commodity production, reducing the supply, hence causing prices of consumer products to rise. This effect stimulates the design and creation of alternative products. Instead of making houses primarily from wood, for example, alternative construction materials from straw bales to recycled waste might become more attractive substitutes. Finally, the remaining timber producers would receive higher profits, making it economically viable to apply environmentally sound forestry practices to timber production.

How would this translate to the western landscape? Using Montana again as an example, most of the state's population resides in a few medium to large cities. This is a trend that is continuing as smaller, rural areas become depopulated. As large portions of the landscape are depopulated, opportunities for restoration are created—from restoration of heavily logged regions in the Cabinet–Selkirks region to reestablishment of bison and grizzlies on the Great Plains in the central part of the state. If we diverted the same amount of money we are currently pouring into rural communities in the form of subsidies, we could easily purchase, fee-simple, millions of additional acres in Montana and incorporate them in large public holdings that would enable us to design biologically sustainable ecosystems.

A similar strategy in western Montana also holds promise for ecosystem restoration. Much of this heavily timbered and mountainous landscape is already in public ownership. What is needed here is termination or significant reduction of destructive extractive industries like livestock grazing and timber harvesting, combined with acquisition of critical private lands that could serve as corridors between major wildlands complexes like the Bob Marshall Wilderness, Selway Bitterroot, and Greater Yellowstone.

Such a program, even if much of the remaining private lands were developed, would result in a number of large ecosystem complexes that are big enough to sustain themselves indefinitely. While it's important to seek development of environmentally friendly extractive methods, we should not lose sight of the need for large, landscape-wide ecosystem preserves. This is the only way we can assure long-term survival of both wilderness and wildland ecosystems into the future.

Perhaps someday we will be in a position to restore part of the midwestern prairies and southern pine forests. But at least we are already in a position to restore the north woods, Buffalo Commons, southwestern desert, and western mountain ecosystems—since we don't at present need to utilize these lands for resource exploitation. A truly "wise use" would be to reestablish viable wildland complexes.

Such a vision has other economic benefits. An American Serengeti could sustain communities that would function as service centers for tourism-based economies. But more than that, these vast landscapes would serve as an ecological bank where species diversity and evolutionary processes could continue. Their value for production of clean water, soil protection and formation, and biological diversity preservation should not be underestimated.

To truly protect wildland ecosystems will require a new commitment to landscape restoration and a significant expansion of our preserve systems. This is far more revolutionary than sustainable development and more ambitious than most conservation organizations are ready to embrace. Yet it is a conservative act: seeking to protect natural landscapes in sufficient quantity and quality to ensure long-term ecosystem viability. Failure to seize the opportunity presented by wildland restoration may result in the failure to maintain not only native ecosystems, but human communities as well.

Minding Wilderness

BILL MCKIBBEN

Complete the following sentence: "You deserve ___ ___ to-day, at ___."

When we talk about wilderness, what do we mean? Not an unaffected place—the deepest wilderness responds deeply to temperature, to rainfall, to a collection of variables so enormous and imponderable that no god could ever keep track of them. And no god would ever need to: the point about wilderness is that it runs on its own, shaped by the appropriate variables, the "natural" ones. A wilderness can change its character over time, usually slowly, occasionally with the speed of a volcanic eruption or a wildfire or flood. It may or may not tend toward complexity or beauty, only toward *fulfillment*, toward expressing its actual nature at any moment—its nature unaltered by the one verboten variable, man. Or at least advanced man: hunter-gatherer man might be okay. It's a little hard to draw the line, but we all know, more or less, where it lies.

I want to attempt an analogy here, not a perfect analogy but perhaps a useful one. The least wild places on earth—that is to say, the places most altered by inappropriate variables, the places tending least toward some ideal expression of their nature—are stripmined hillsides, plantation forests, shopping center parking lots, and the insides of human heads. Our minds, buzzed every waking second by jingle–newscast–nature documentary–music video–college football scoreboard–magazine article–radio talk show–investment advice–weather forecast–hamburger commercial–mail order catalog–environmental anthology–video rental–Walkman, turn into paved-over sacrifice zones where exotics choke the few native species trying gamely to push on up.

You deserve a break today, at McDonald's. That sentiment is on a billboard in our brains, one of a hundred thousand billboards that obscure our view until we gradually forget there is a view.

A grand analogy, perhaps, but its defect is obvious: the "natural" state of the human mind, the hunter-gatherer mind, say, is not free from human manipulation. It is mostly the *product* of human manipulation plus whatever weight you give to genetics and deities. The human mind is a human construct. But a human construct of a certain sort, pointed in definite directions. Human societies, especially tribal societies, used to give enormous attention to the shaping of minds—not approaching the task in any haphazard fashion but making it, rather, the focus of ritual. Certain ideas, usually ideas that allowed people to live in a particular place for an infinite time, were passed carefully down in story, cult, and scripture. The test of these ideas was their practical fitness for society as a whole.

For better and for worse, the progress of modernity has been the breakdown of these systems, the substitution of other tests and hence other educations. What we call history is the rise and fall of various schemes for shaping the mind, a process that has been gradually speeding up until, in our age, it is a blur. Judging from the last few centuries, and particularly the current evidence, the direction of that change has been clear: we are schooled in intense individualism and short-term gratification. *You, a break, today.* This education, though it pretends "freedom" and "choice," is at least as effective as any Plains Indian rite or South Seas coming-of-age. We stare into our video campfire, and it shines back from our eyes.

But this new indoctrination, I would argue, represents "unnatural" influences on the wilderness of our minds, as opposed to the "natural" ones of a humbler era. It is not only commerce that forces this worldview on us, though that is the strongest force. Literature and art too—the stories of our time, repeated ceaselessly in sitcom, music video, and big novel—are about overthrowing old conformities in the name of "liberation." And of course most of those old conformities needed overthrowing—father rarely knew best. But all our liberation *from* has yet to yield a liberation *for*. The idea that we will live without some vision in our minds is illusory. Left unguarded, our minds fill with the commands of consumer society, with the selfishness, sleaze, and product-lust that are the cheatweed, purple loosestrife, and cowbirds of our culture.

Nazis thought like this, of course, and Stalinists bent on making a new man. Their butchery haunts anyone even thinking of prescribing a "common vision" for people. Given the choice of keeping what we have right now or adopting the prescriptions of, say, the religious right, I'd shut up. Much better an unprotected, eroding wilderness ringed by billboards than one chopped down and replanted with "superior" trees. But the power of environmental analysis is twofold. First, it lets us see the high practical cost of continuing on our present course. Perhaps more important, its worldview suggests a series of small-scale, decentralized visions, a third route between our current anarchy and totalitarianism.

———

I am reasonably sure there is a state to which the human mind would tend at this stage in our evolution, just as there is a direction in which hardwood forests at a certain altitude and soil type and slope will tend given certain climatic conditions. Tendency is not monolithic, of course, but on the whole, it seems to me, we were built for close community and service, some contact with the natural world and its cycles, a certain amount of physical labor, and perhaps some sense of the divine. My evidence is both flimsy and iron: everyone I know well enough to discuss these questions responds they feel most *right* when expressing their nature in such situations. And that is all the more remarkable because these are the very states the 150-channel TV system, the shopping mall, and the suburb try most diligently to ridicule and ignore. With their endless broadcasts TV can jam the signals we would otherwise hear. But when we escape its buzz—climb a mountain, volunteer in a homeless shelter, pick up a guitar—we are quickly reminded where real satisfaction lies.

To extend this analogy further: just as we try to fence off wilderness, to keep it untrammeled by that which would alter it unnaturally, so we would profit by protecting our minds and the minds of those we love. It is not easy. A few years ago, writing a book, I watched every program that came over the nation's largest cable system for a single day—some two thousand hours of tape. It was destructive not only because of the emotions it aroused—dissatisfaction, envy, and fear—but because it deployed irony as a tool to keep you from getting upset. TV, the whole consumer culture, wears David Letterman's grin: "Don't bother calling this silly and manipulative. Everyone *knows* it's silly and manipulative. Just watch it." You feel like a fool fighting back. And yet, unresisted, it changes you. What might have been an old-growth forest up there turns into clearcuts and paper birch. I haven't had a TV for years, and I don't live in a suburb, but I know my all-American upbringing still rules an astonishingly large part of my heart. I suppose I'm engaged in a process of wilderness restoration. Not in *my* lifetime . . . but I have a baby daughter now.

Protecting our minds is not the only task, of course. We need also to recover and invent the stories, myths, aspirations, and games that bend our heads in a more "natural" direction, toward satisfaction and stasis instead of growth. This work seems to me part and parcel of protecting real wilderness, the unmetaphorical mountains and forests and reefs. The threats they face—abusive logging, condos, ski resorts, the even more powerful onslaught of global forces like the greenhouse effect—are products of souls steeped in convenience, comfort, acquisition, and not in connection, meaning, and contentment. Legislation, lobbying, tree-sitting—all are vital. But at their truest, they come from deep places in us and draw on deep places in others. Deep places that need protection.

Covers the Ground

GARY SNYDER

"*When California was wild, it was one sweet bee-garden . . .*"

Down the Great Central valley's
blossoming almond orchard acres
lines of tree-trunks shoot a glance through
    as the rows flash by—

And the ground is covered with
cement culverts standing on end,
house-high & six feet wide
culvert after culvert as far as you can see
    covered with
mobile homes, pint size portable housing, johnny-on-the-spots,
concrete freeway, overpass, underpass,
    exit floreals, entrance curtsies, railroad bridge,
long straight miles of divider oleanders;
scrappy ratty grass and thistle, tumbled barn, another age,

yards of tractors, combines lined up—
new bright-painted units down at one end,
old stuff broke and smashed down at the other,

cypress tree spires, frizzy lonely palm tree,
steep and gleaming
fertilizer tank towers fine-line catwalk in the sky—
	covered with walnut orchard acreage
irrigated, pruned and trimmed;
with palleted stacks of cement bricks
	waiting for yellow fork trucks

quarter acre stacks of wornout car tires,
dust clouds blowing off the new plowed fields,
taut-strung vineyards trimmed out even on the top,

cubic blocks of fresh fruit loading boxes,
long aluminum automated chicken feeder houses,
	spring furz of green weed
	comes on last fall's baked ground,
		"Blue Diamond Almonds" farther see
identical red-roofed houses closed-in fencing,
stretching off towards the tower that holds catfood
with a red / white checkered sign

crows whuff over almond blossoms
beehives sit tight between fruit tree ranks
eucalyptus boughs shimmer in the wind—a pale blue hip-roof house
	behind a weathered fence—

crows in the almonds   trucks on the freeways
Kenworth, Peterbilt, Mack
rumble diesel depths,
like boulders bumping in an outwash glacial river

drumming to a not-so ancient text

*The Great Central Plain of California
was one smooth bed of honey-bloom
 400 miles, your foot would press
a hundred flowers at every step
it seemed one sheet of plant gold;*

> *all the ground was covered
> with radiant corollas ankle-deep:
> bahia, madia, madaria, burielia,
> chrysopsis, grindelia,
>> wherever a bee might fly—*

us and our stuff just covering the ground.

Said one John Muir.

This poem is part of a work in progress entitled "Mountains and Rivers Without End."

DISPATCHES

There does exist a possibility that we can live more or less in harmony with our native wilderness; I am betting my life that such a harmony is possible. But I do not believe that it can be achieved simply or easily or that it can ever be perfect, and I am certain that it can never be made, once and for all, but it is the forever unfinished lifework of our species.

—*Wendell Berry,* Home Economics

Over the past three decades the focus of conservation activism has evolved from attempts to save a few big trees by creating reserves and wilderness enclaves, to efforts to protect bellwether species such as salmon and the spotted owl, to a broader, more encompassing concern for the restoration of wildlands and the preservation of vast intact ecosystems and biological diversity.

The contributors to this section are some of the key players in a new movement, the drummers who daily search for ways to call our attention to diminishing wild places. Their words are not the cries of oracles of doom, for their dispatches carry a note of hope and prospects for renewal. They speak of the types of maps and tools we need to begin the process of reaching a rapprochement *between the needs of wildlands and the needs of human culture. As E. O. Wilson has written in* The Diversity of Life, *"let us go beyond mere salvage to begin the restoration of natural environments, in order to enlarge wild populations and stanch the hemorrhaging of biological wealth. There can be no purpose more enspiriting than to begin the age of restoration, reweaving the wondrous diversity of life that still surrounds us."*

The chorus of voices in this concluding section send out a challenge to each of us. Collectively they admonish us to examine our motives and our actions and urge us to review our cherished beliefs about wildlands. By resetting the debate in new terms, with a new and expanded focus, they are pointing the way toward significant changes in the prospects for both biological and human diversity.

Barry Lopez, in his essay "The Rediscovery of North America," provides some perspective:

> *What we need is to discover the continent again. We need to see the land with a less acquisitive frame of mind. We need to sojourn in it again, to discover the lineaments of cooperation with it. We need to discover the difference between the kind of independence that is a desire to be responsible to no one but the self . . . and the independence that means the assumption of responsibility in society, the independence of people who no longer need to be supervised. We need to be more discerning about the sources of wealth. And we need to find within ourselves, and nurture, a profound courtesy, an unalloyed honesty.*

Where Man Is a Visitor

DAVE FOREMAN

In past years, opposition to Wilderness Areas and National Parks came from economic interests who wished to exploit publicly owned natural resources, often with government subsidies. It still does. More recently, opposition has also risen from an ideological stand of People First! Edward Abbey is often quoted as saying, "Wilderness needs no defense, only defenders." Cactus Ed told me he never said that and it didn't make sense to him. Clearly, wildlands need defenders. But with an ideological attack by prideful humanists,[1] Wilderness needs to be defended as both concept and place.[2]

Wilderness defenders from the fields of anthropology, archaeology, history, rural sociology, philosophy, and conservation biology need to investigate the claims of the humanist opponents of Wilderness and rebut them. And all lovers of things wild and free need to articulate a ringing defense of Wilderness. What follows is neither a comprehensive defense nor a fully explored rebuttal. Because of that, it is largely unreferenced. It is a map scratched in river sand with a stick of driftwood. I hope it leads some of us to tease out deeper answers from the questions raised here.

One hundred and forty years ago, sometime wilderness canoeist, sometime suburban bean farmer, Henry David Thoreau said, "I wish to speak a word for Nature, for absolute freedom and wildness." Today, I wish to speak a word for Nature With Integrity, for Wilderness Areas.

The opening section of the 1964 Wilderness Act defines Wilderness in part as an area "where man himself is a visitor who does not remain." In recent years, both charlatans and those sincerely concerned with the estrangement of humans from nature have seized on the notion of Wilderness as a place where humans are visitors. They use the lack of human habitation to chide Wilderness Areas as misanthropic and self-defeating. These scolds come from around the globe and from all over the political map.

We can dismiss from this discussion the hairy-chested populism calling itself the "wise-use movement." These largely rural loggers, miners, ranchers, dirt bikers, road hunters, and trappers spit on Wilderness because it "locks them out." They seem to have evolved into a new species without walking ability. Perhaps we can name them *Homo petroliens*. Their opposition to Wilderness and other melodies of conservation and environmentalism stems from a know-nothingism that wears anti-intellectualism and anti-elitism like a Stihl chainsaw cap. They are the handmaidens of America's most elite: the corporate oligarchy. These rugged individualists are puppets on strings manipulated by oil, mining, and timber executives. They offer no serious critique of Wilderness.

More troubling are those critics of Wilderness Areas who hail from more progressive and thoughtful traditions in America and abroad. Some are even conservationists, biologists, and ecophilosophers. Others are environmentalists who I suspect have never been outside except to hail a cab. Let me try to round up some of their arguments against Wilderness, and then blow a few smoke rings in response.

I don't wish to be crotchety in what follows, but I am tired of anti-Wilderness nagging, so I will be short. I respect many of the folks whose recent writings question traditional notions of Wilderness Areas and National Parks. I think I understand their frustrations and hope my scratching in the sand will lead to confluence instead of divergence. Because I do not wish to fall into the trap of demonizing others (a popular pastime in academic philosophy—see the pages of *Environmental Ethics*), I will generally not attach the names of individuals to the anti-Wilderness arguments that follow. Somewhat different anti-Wilderness arguments are emphasized in Latin America, Asia, and Africa than in the United States, Canada, and Australia. Yet, while there are various hybrids, they are a single herd.

I summarize the arguments against the concept of Wilderness as follows:

- Wilderness Areas separate humans from nature because people are not allowed to live in them.
- What needs to be protected is wildness, not Wilderness. Wildness is real; Wilderness is a human construct.
- Wilderness Areas and their kin, National Parks, are a legacy of Western civilization and its false dichotomies.
- Wilderness supporters are misanthropic. Not only do we seek solitude in unruly places where no person lives, but we wish to exclude indigenous people and rural dwellers from the land.
- The environmental movement should not concern itself so much with wildlife and parks. Environmentalism is fundamentally about human health.
- We should encourage people to reinhabit the land. National Forests and other public lands should be open to settlement by people wishing to be caretakers and land healers.
- Indigenes (and even rural dwellers in North America) improve on nature with their activities. Because of their deep knowledge of the land and love for it, their activities increase biological diversity.
- The notion of a pristine pre-Columbian America is a myth. Native peoples greatly modified the land. The paradise Europeans encountered was not a wilderness but a garden improved by humans.
- Belief in the intrinsic value of other species and natural processes is an affectation and luxury unique to wealthy Westerners. Such notions as manifested in National Parks and Wilderness Areas are foreign to the Third World. People in "developing" countries need sustainable development that will put food in their bellies and shoes on their feet—not reserves for wild animals. The idea of Wilderness is alien to them. Wilderness Areas and National Parks are legacies of European and U.S. imperialism.
- The role of humans is to garden the Earth. We must seize the tiller of the planet and actively manage evolution and natural processes. Wilderness Areas represent avoidance of responsibility and dereliction of duty. Without active management and hu-

man presence, natural landscapes will deteriorate and lose biodiversity.

- At over 5 billion people and growing, we can't afford Wilderness Areas or large carnivores. The land and ocean have to produce for people. In a world rife with economic imperialism and social injustice, progressives must direct their efforts to improve the lot of all people. We can't waste time on lions and tigers and bears or on outdoor museums and backpacking parks for the economic elite.

- Changing demographics in the United States mean that traditional support for Wilderness Areas will decline. Conservationists need to refocus on more people-oriented environmental issues to bring Hispanics, Asians, and African-Americans into the movement.

- Scientists tell us that Earth is in the throes of the sixth great extinction episode. Wilderness Areas and National Parks have therefore failed to protect biodiversity. Something new is needed: "ecosystem management" that includes people.

Whew. This begins to look like an FBI indictment that piles on everything except failing to floss each night. Nor have I exhausted the anti-Wilderness arguments of sustainable development advocates and environmental humanists.

Now, arguments do not materialize out of thin air. They come from people. These people have motivations for clamping their teeth on certain viewpoints. Once upon a time, wicked little boys liked to pour turpentine on cats to watch them squall and run. Some will think my concern about motivation for anti-Wilderness agitation is splashing turpentine. I may be a little cranky, but, honestly, I'm basically a nice guy. Ask my sister. I'm more interested in discovering common ground than in building fences. Yet motivation is important. I will consider motivation as I try to point in the direction of possible answers to anti-Wilderness criticism.

I fear that some of the most vociferous critics of American-style Wilderness and National Parks are suffering from Third World jingoism. Wilderness is a victim of chronic anti-Americanism. Everything from the United States is bad to some folks. North America and Europe are to blame for all the world's problems.

Some from the United States are expiating white liberal guilt. (I'm lucky. I come from, at best, a lower-middle-class family of Scotch-Irish hillbillies. I have more than my share of moral failings, and one of these days I reckon I should get around to atoning for some of them, but guilt for being pampered I have not.)

Western civilization (imperialism) and the United States of America deserve plenty of criticism. And I think the United States should be held to higher standards than any other country or society because we have claimed from the beginning to be engaged in a superior social experiment. I have personally experienced some of America's failings. (When your government spends three million dollars to try to frame you, even a slow guy like me develops a bit of skepticism.)

But the United States is not wholly evil. We are not the sole source of injustice in the world. Despite the efforts of J. Edgar Hoover and Ronald Reagan, we have a Bill of Rights and we jealously guard it. For all of our lapses, no other nation past or present can come close to the United States on protection of civil liberties. The Bill of Rights is recognized as the United States' great gift to the world. And we have given the world an even greater gift: the idea of National Parks and Wilderness Areas.

Moreover, the anti-Americanism inherent in Third World criticism of Wilderness and National Parks ignores the venality of elites in those countries. To blame white males for all the world's problems is—dare I say it?—racism. Furthermore, the leading Third World critics of Wilderness are Western-educated members of the economic/social elite in their own countries.

None of this is to argue that we should ignore issues of international economic justice. Europe, Japan, and the United States, in cahoots with the robber barons of the Third World, wage conscious economic imperialism against people around the world. Furthermore, we need to safeguard land for use by indigenous peoples and peasants and to recognize and celebrate their knowledge and stewardship of the land. Wilderness Areas and National Parks need not conflict with the needs and rights of the downtrodden.

Several anti-Wilderness arguments come from the myth of the Noble Savage. Alienated from our own "corrupt" society, we still want to believe that humans are intrinsically good, so we romanticize indigenous peoples as the first ecologists. It seems we can't accept nonindustrial societies for what they are. We have to demonize them as savages with animal lusts and an incapacity for civilization, or we have to exalt them as paragons of virtue.

Anthropology is like the Bible: you can use it to support any claim about humans and nature you wish. We can argue until we're blue in the face about the level of impact indigenous people had in the Americas, for example. The wisdom until recently was that Native Americans had very little effect on the landscape. New England's Puritans argued so to justify their taking of "unused" land from the Indians. The pendulum has swung the other way in recent years. Now crackpots as well as serious scholars claim that even small populations significantly altered pre-Columbian ecosystems—especially through burning. The "myth of pristine America" is in disrepute. Worshipers of the Noble Savage argue this impact was positive. Some even place the Aztecs, Incas, and Mayans on the ecological pedestal, too.

Many researchers, however, see evidence of ecological collapse in archaeology. Did the Hohokam and Anasazi of the American Southwest overshoot carrying capacity and cause ecosystem failure? Newly mobilized with Spanish mustangs, would the Plains Indians have caused near extinction of the bison had they been left alone for another hundred years? Did the civilizations of Mesoamerica degrade their lands as terribly as the civilizations of the Middle East and Mediterranean? Was the extinction of the Pleistocene megafauna caused by Stone Age hunters entering virgin territory?

Deeper questions follow. Is the profound philosophy of the Hopi a result of a new covenant with the land following the Anasazi ecological collapse seven hundred years ago? Could the hunting ethics of tribes in America (and elsewhere) have been a reaction to Pleistocene overkill?

Where some argue that shifting human horticulture and using native plants increases local species diversity, dare we ask about the quality of that increased diversity? Are the additional species common, weedy ones? Are many of these exotics from Europe? All biodiversity is not equal. Rare, sensitive, native species are more important than weeds which do well in human-disturbed areas.

In certain areas of the Americas, high population density and intensive agriculture led to severely degraded ecosystems. The first wave of skilled hunters twelve thousand years ago probably caused the extinction of dozens of species of large mammals inexperienced with such a predator. But I question, as some argue, that the North American forests and prairies found by the first European explorers and colonists were primarily the result of burning by native tribes. Certainly in localized areas North American tribes had a major impact on vegetation because of anthropogenic burning. But how extensive could this manipulation have been with a population of only 4 to 8 million north of the Rio Grande in 1500?[3]

These questions are not to oppose the legitimate land claims of Native Americans and other First Nations. In most cases, tribes are better caretakers of the land than government agencies. Despite the opposition of other conservationists in New Mexico, I supported the transfer of Carson National Forest's Blue Lake area to Taos Pueblo in the 1970s. It was their land and they have done a far better job of protecting its wilderness than the Forest Service would have. But we must be intellectually honest in investigating human relationships with the land, and we must not pander to the Noble Savage myth and then hold primal peoples up to impossible standards.

It is a far leap from holding up indigenous people or Third World peasants as ecologists to making the same claim for rural residents of the United States. I have found an abysmal lack of ecological knowledge or appreciation among America's bumpkin proletariat. I've asked ranchers to identify plants for me. They know winterfat and a few species of grass, but most else is "weeds" or "brush." Similarly, loggers and even forest rangers know the trees valuable for timber, but shrug their shoulders at the rest. Attitudes to animals are even more appalling. My neighbors in Catron County, New Mexico, scoffed at university biologists. I was often told that spiny lizards were baby Gila monsters.

Notwithstanding the seesawing over preindustrial societies' role in changing the face of the earth, there is much evidence that Wilderness—vast tracts uninhabited by humans—is not an alien concept to primal cultures. Native Hawaiians tell me that before the American conquest, some mountains were forbidden to human entry—on pain of death. Jim Tolisano, an ecologist who has worked for the United Nations in many countries, reports that the tribes of Papua New Guinea zone large areas off-limits to villages, horticulture, hunting, and even visitation. "You don't go there; that mountain belongs to the spirits." Like the Papuans, the Yanomami of the Amazon engage in fierce blood feuds (my hillbilly ancestors in eastern Kentucky were a lot like them). Between villages is a death zone where one risks one's life by entering. As a result, large areas are left uncultivated, unhunted, and seldom visited. Biodiversity thrives there. These borderlands are refugia for the animals intensively hunted near settlements.[4]

Some anthropologists think that the permanent state of war between some tribes is an adaptation to prevent overshooting carrying capacity, which would result in ecological collapse. (These unused areas on territorial borders are uncannily similar to the places where the territories of wolf packs abut one another and deer occur in high densities.) My forebears were able to follow Dan'l Boone into the "dark and bloody ground" of Kaintuck because it was uninhabited by Native Americans. The Shawnee north of the Ohio River and the Cherokee from the Tennessee Valley hunted and fought in Kentucky. But none lived there. Wasn't it a wilderness area until the Scotch-Irish from Shenandoah invaded?

———

Geographers, anthropologists, historians, and ecologists need to research these tantalizing threads and others to show that Wilderness Areas—where humans are visitors who do not remain—were once widespread throughout the world. Wilderness Areas are not a uniquely twentieth-century idea of Americans, Canadians, and Australians.

Can people outside the English frontier colonies appreciate wilderness for its own sake? I know Native Americans and have met folks in Mexico and Belize who are just as supportive of Wilderness as I am and who believe in the intrinsic value of other species. At an international wilderness mapping conference a few years ago, I met South American biologists as intransigent in their defense of Wilderness as Reed Noss. Jim Tolisano tells of colleagues in Sri Lanka, several African countries, Costa Rica, and the Caribbean who make me look like a weenie.

It is racist to claim that only middle-class North Americans and Australians can be deep ecologists or Wilderness supporters. How dare pampered, pompous twits claim non-Westerners are incapable of a land ethic like Leopold's? Any claim that the people of Latin America, Africa, and Asia can live without wild things and sunsets is another form of imperialism. It is just as racist to claim that African-Americans and other people of color in the United States are not interested in wildlife and Wilderness. Michael Fischer, former executive director of the Sierra Club, tells a story of being on a radio program with Ben Chavis, now executive director of the NAACP. Fischer wanted to emphasize the Sierra Club's new willingness to concern itself with issues of environmental justice in minority communities and suggested that the Sierra Club should focus less on Wilderness issues. Chavis reproved him for pandering. "We're interested in wildlife and wilderness, too." Wilderness conservationists need to concern themselves with pollution issues in communities of color. We also need to help inner-city residents gain wilderness experiences. We do not need to soften our resolve to protect Wilderness.

There is a sincere and valid belief that we need to create a new, bioregional, sustainable society of people on the land. Caretaking and restoration on a watershed basis are the real work. Gary Snyder tells a beautiful story of walking down the trail to his home at dusk. He surprised a mountain lion sitting outside the window listening to his stepdaughter play the piano. We need such communities as San Juan Ridge in California's Sierra Nevada foothills. Arne Naess calls them "mixed communities." But extensive Wildernesses where humans only visit are also essential. Large carnivores and many other species need refugia away from constant human presence. We humans do not always make good neighbors, and the Gary Snyders are rare.

Wilderness Areas and National Parks have indeed failed to safeguard the full range of native species and ecosystem functions in North America. But is it the idea of Wilderness Areas and National Parks that has failed, or is it rather the application in our politicized arena of land management that has failed? Because of the superior political strength of extractive industries, conservationists have been defeated time and again as they have tried to establish protected areas. The biologically more productive areas of our federal lands have been released for clear-cutting, dam building, road building, mining, and vehicle play.

I have spent many, many days and nights in Wilderness Areas from Alaska to Central America. I have not found that these landscapes where I am only a visitor separate me from nature. When I am back-packing or canoeing in a Wilderness, I am home. Wilderness is the reality; wildness is a human concept.

Aldo Leopold wrote that there are those who can live without wild things and sunsets, and those who cannot. Many of those who criticize Wilderness Areas show no gut passion for wild things. Do they hear goose music or thrill to the first pasqueflower? Do they hunger to know there are wolves hunting moose in a place untrammeled by humans? I see no evidence of such feelings in the writings of some Wilderness critics.

A few suffer from what I will politely call the Little Red Riding Hood complex. They're scared of the dark. They're terrified of the big bad wolf. Vest-pocket nature preserves for rare plants are fine. To them, that's what the preservation of biodiversity is all about. But big, wild, uninhabited places scare the bejeesus out of them.

We can read history as the progressive control of Earth by Lord Man. Those New Age technocrats who prattle about seizing the tiller of evolution and *improving* Earth are then monstrous heirs of Greek hubris. Their lovely human garden is hell for other species. Four billion years of life becomes a mere overture, before Man, in all his Wagnerian glory, strides singing onto the set. Does our madness know no bounds? Have we no humility?

———

Enough counterarguments. Wilderness Areas, more than anything else, challenge us to be better people.

For six thousand years, each succeeding age has puffed out its chest more. As each Ozymandias falls to the lone and level sands, a greater and more prideful Ozymandias takes his place. Virtue is replaced more and more by might and the will to power. Wilderness Areas are the quiet acknowledgment that we are not gods.

Wilderness Areas where humans are visitors who do not remain test us as nothing else can. No other places teach us humility so well—whether we go to them or not. Wilderness asks: Can we show self-restraint to leave some places alone? Can we consciously choose to share the land with those species who do not tolerate us well? Can we develop the generosity of spirit, the greatness of heart, not to be everywhere?

No other challenge calls for self-restraint, generosity, and humility more than Wilderness preservation. Our Concord friend, Thoreau, said: "In Wildness is the Preservation of the World." True, so true. But a deeper truth is that in Wilderness Areas is the Preservation of Wildness.

Notes

1. I use "humanist" in the sense of those who glorify Lord Man and who are outwardly anthropocentric on conservation issues. See David Ehrenfeld's *The Arrogance of Humanism* (New York: Oxford University Press, 1978).
2. Where Wilderness refers to Wilderness Areas as concept or as on-the-ground designation, I capitalize it. Much of what follows also applies to National Parks.
3. The best recent estimates of serious demographers.
4. George Schaller reports that when Amazonian tribes were armed only with blowguns and bows, monkeys could be found half a mile from villages. Now, with the advent of the shotgun, monkeys are not found within five miles of settlements. Jim Tolisano reports similar changes in Papua, New Guinea.

A Sidelong Glance at
The Wildlands Project

JOHN DAVIS

Metaphorically and literally, this anthology bespeaks The Wildlands Project—the North American wilderness recovery strategy.[1] Metaphorically speaking, these essays, poems, and introductions are like the cores, buffer zones, and corridors being designed by The Wildlands Project. Like the ecological reserve system envisoned by the project, this anthology will enhance exchange, movement, and migration—only, in this book the movers are ideas and writers rather than genes and species. Literally speaking, this book was inspired by The Wildlands Project and addresses many of the tough questions implicitly raised by the concept of continental wilderness recovery.

Wild Earth, the regular voice of The Wildlands Project, cannot in its overcommitted and overstuffed pages even begin to do justice to the many sociological and political questions ancillary to any discussion of rewilding the continent. *Wild Earth* stresses biology. The need was clear for a forum where players would discuss not so much nature per se (*Wild Earth*'s main bent) but the human/nature interface—hence this anthology. The book would not be complete, though, without a basic explanation of The Wildlands Project. Other more skilled essayists have poked, prodded, and pondered the underlying notion that all Natives ought to be allowed their place and space on Turtle Island. As *Wild Earth* editor and Wildlands Project board member, it is my task to describe the premises upon which the project is based.

Fearing, however, that North American wilderness recovery efforts could easily be diluted by calls for moderation and political compromise (as happens to most radical endeavors), I wish to go a bit beyond a basic outline of the project's presuppositions and suggest more radical ideas consistent with North American wilderness recovery. To avoid defaming fellow Wildlands Project participants, I'll venture afield only after attempting to delineate the common ground. That is, I'll first suggest the assumptions shared by practically all wildland proponents and then offer some radical notions that at least a few wildland people hold.

CONSENSUS PRESUPPOSITIONS

Let me first quote the best concise statement yet offered of The Wildlands Project's aims. While the project's mission statement (see appendix at the end of this essay) offers an eloquent explanation of what North America needs, Reed Noss in the same issue (*Wild Earth*'s special issue on The Wildlands Project, 1992) offers four ecological goals that really ought to be emblazoned above the doors of every land management agency in the country:

1. Represent, in a system of protected areas, all native ecosystem types and seral stages across their natural range of variation
2. Maintain viable populations of all native species in natural patterns of abundance and distribution
3. Maintain ecological and evolutionary processes, such as disturbance regimes, hydrological processes, nutrient cycles, and biotic interactions, including predation
4. Design and manage the system to be responsive to short-term and long-term environmental change and to maintain the evolutionary potential of lineages

Wildland proponents presuppose these goals. The following premises, though awkwardly worded, would also find agreement among wildland advocates:

1. Anthropocentrism is ethically repugnant and biologically indefensible. All creatures have equal rights to their natural places on earth. Humans need to adopt an ecocentric paradigm and cease dominating the world.

2. Humans therefore have an obligation to allow all natural species and processes to fill their natural places and fulfill their natural functions. That means satisfying the aforementioned goals. It also means reducing human population (through such benign measures as education and universal availability of birth control) and turning much of the landscape back over to natural forces (as opposed to industrial and technological forces).

3. The wildlife reserve systems on this continent, and all others, are grossly inadequate. Designated reserves are too small and isolated to maintain viable—let alone naturally fluctuating—populations of all native species.

4. Far too much wildlife habitat has been destroyed. Stopping habitat destruction is not enough. We must also commence ecological restoration. In some places, this will entail simply removing human obstructions (such as roads and dams) and letting nature be. In other places, it will entail active restoration; which may include reintroducing species, prescribing burns, planting trees, shoring up eroding stream banks, and other labor-intensive techniques.

5. We need to design a continental ecological reserve system consisting of large wild core areas surrounded by buffer zones and linked by habitat corridors to restore and protect biodiversity. This will be a long-term effort as the reserve system grows through time. Again, concurrent with this, we must work to lower human population through benign means.

All these premises have been adequately defended by deep ecologists and conservation biologists elsewhere. The Wildlands Project work engendered by these premises has also been discussed at length, especially in *Wild Earth*'s special issue. At this stage, wildland proponents are working within their regions to formulate wilderness proposals which will describe with maps and text the reserves needed to restore and protect biodiversity. As they are completed, these proposals will be published in *Wild Earth* or as special publications. When the whole continent is covered, they will be published in a large-format book: a North American wilderness recovery strategy that will set the conservation agenda for decades to come.

MINORITY SUPPOSITIONS

Others will strongly disagree, but some wildland advocates believe that the vast bulk of North America must be allowed to become wild again if we are to achieve the four basic Wildlands Project goals. Those who disagree will have a particularly hard time with Reed Noss's wonderful phrase "in natural patterns of abundance and distribution." Sixty-five million bison, for instance, will not abide extensive cornfields. And this raises the first minority assumption.

Assumption 1: We need to be very clear about scale, both spatial and temporal. Spatially speaking, we are probably being less than honest if we imply that North America's biodiversity can be saved in a system of reserves covering a minority of this continent's acreage. For what little it's worth, I venture to suggest that fulfilling the basic four goals would require that at least 90 percent of the continent as a whole (including ocean waters to the edge of the continental shelf) be protected as wild habitat as soon as possible. Of course, in some heavily developed regions, reserves will be small initially, but they can and must expand.

In temporal terms, we may leave people a bit befuddled when we talk about ours being a project of centuries. Yes, ensuring the opportunity for long-term prosperity of all (remaining) native species and processes will entail centuries of work (and nonwork), but much could be accomplished quickly. Most dramatically, the United States could secure well over half a billion acres of potential wilderness and actual wilderness with a few pen strokes (of the president and some governors) by banning commodity extraction and motorized use of public lands. We could save over 10 million sparsely inhabited acres in the northern forests by allocating a few billion dollars of federal money (the cost of a few Stealth bombers) to purchase lands from timber companies (which are trying to pull out anyway). Canada offers similar prospects. Much of the continent remains undeveloped and could be quickly secured if the political will can be mustered.

Assumption 2: We sometimes talk about a future ecological reserve system as if it can be superimposed upon the existing sociopolitical systems of this continent. Any recent road atlas belies such a hope. In the words of Jamie Sayen: "Industrial civilization is incompatible with biodiversity." Quite likely, industrial civilization will eventually collapse of its own accord. It may, then, be counterproductive to alarm people by proclaiming the need to dismantle industrial civilization. Let's not, though, pretend that The Wildlands Project's goals can be reached without systemic changes. The project's role may not include calling for these systemic changes, but let's be prepared to discuss them.

Assumption 3: We should look askance at technology, too. Almost all modern technologies are built and used at the expense of the natural world. In the near term, wildland proponents will rely upon many modern technologies (telephones, computers, cameras, and such) to spread their message; but we should at least entertain the possibility that a future wild world will be free of motors, firearms, electronic gadgets, and the like. Even in the short term, serious questioning of technology could benefit wildlands. In many reserves, wildness could be ensured simply by banning destructive technologies without needing to ask people to relocate. As I argued in *Wild Earth*'s special issue on The Wildlands Project, the problems in the Adirondacks would mostly disappear, and extirpated species would likely reappear, if motors and firearms were banned from the park.

Assumption 4: We should not paint the battle for biodiversity as an all or nothing struggle. Such an approach would cause people to despair and surrender, for the odds of complete victory are infinitesimally small. (One hundred species will go extinct tomorrow if recent estimates are right.) Every acre saved is a victory; every hectare, 2.5 victories.

Assumption 5: It is incumbent upon Americans (especially those of us urging others to change) to simplify their lifestyles, to consume less. North America will never become wild and healthy again so long as Americans maintain their current levels of consumption. Again, The Wildlands Project has enough work without adding efforts to convince people to be frugal, but advocates of the wild should set good examples. Wildland proponents who need to use cars, planes, computers, fax machines, electric hair dryers, and other machines to do their wildlands work will continue to do so. Those who can do without machines should. This fifth point brings up the related need to restore human as well as natural communities. Creating economies based on restoring rather than destroying nature will require healthy self-sufficient human communities.

A FINAL KEY POINT

Wildland proponents agree on another key point: We should employ the tactics and voice the ideas that will save the most habitat. If publicizing unconventional ideas like the ones cited here will only turn people away, then maybe we'd best keep our most radical notions to ourselves. I believe, however, that we underestimate people's biophilia—suppressed but smoldering still, deep within—and the appeal of primeval wildness if we assume we must refrain from speaking hard truths. Intelligent people throughout the world are ready to end the war on wildlife. A paradigm shift is in the air; the North American Wilderness Recovery Strategy is the way to put it on the ground.

APPENDIX: THE WILDLANDS PROJECT MISSION STATEMENT

Our Mission

The mission of The Wildlands Project is to help protect and restore the ecological richness and native biodiversity of North America through the establishment of a connected system of reserves.

As a new millennium begins, society approaches a watershed for wildlife and wilderness. The environment of North America is at risk and an audacious plan is needed for its survival and recovery. Healing the land means reconnecting its parts so that vital flows can be renewed. The land has given much to us; now it is time to give something back— to begin to allow nature to come out of hiding and to restore the links that will sustain both wilderness and the spirit of future human generations.

The idea is simple. To stem the disappearance of wildlife and wilderness we must allow the recovery of whole ecosystems and landscapes in every region of North America. Allowing these systems to recover requires a long-term master plan.

A feature of this design is that it rests on the spirit of social responsibility that has built so many great institutions in the past. Jobs will be created, not lost; land will be given freely, not taken.

Our Vision

Our vision is simple: we live for the day when Grizzlies in Chihuahua have an unbroken connection to Grizzlies in Alaska; when Gray Wolf populations are continuous from New Mexico to Greenland; when vast unbroken forests and flowing plains again thrive and support pre-Columbian populations of plants and animals; when humans dwell with respect, harmony, and affection for the land; when we come to live no longer as strangers and aliens on this continent.

Our vision is continental: from Panama and the Caribbean to Alaska and Greenland, from the Arctic to the continental shelves, we seek to bring together conservationists, ecologists, indigenous peoples, and others to protect and restore evolutionary processes and biodiversity. We seek to assist other conservation organizations, and to develop cooperative relationships with activists and grass-roots groups everywhere who are committed to these goals.

The Problem

We are called to our task by the failure of existing Wilderness, Parks, and Wildlife Refuges to adequately protect life in North America. While these areas preserve landscapes of spectacular scenery and areas ideally suited to non-mechanized forms of recreation, they are too small, too isolated, and represent too few types of ecosystems to perpetuate the biodiversity of the continent. Despite the establishment of Parks and other reserves from Canada to Central America, true wilderness and wilderness-dependent species are in precipitous decline:

- Large predators like the Grizzly Bear, Gray Wolf, Wolverine, Puma, Jaguar, Green Sea Turtle, and American Crocodile have been exterminated from most of their pre-Columbian range and are imperiled in much of their remaining habitat. Populations of many songbirds are crashing and waterfowl and shorebird populations are reaching new lows.
- Native forests have been extensively cleared, leaving only scattered remnants of most forest types. Even extensive forest types, such as Boreal Forest, face imminent destruction in many areas.
- Tall Grass and Short Grass Prairies, once the habitat of the most spectacular large mammal concentrations on the continent, have been almost entirely destroyed or domesticated.

The Meaning of Wilderness

The failure of reserves to prevent the losses just mentioned rests in large part with their historic purpose and design: to protect scenery and recreation or to create outdoor zoos. The Wildlands Project, in contrast, calls for reserves established to protect wild habitat, biodiversity, ecological integrity, ecological services, and evolutionary processes—that is, vast interconnected areas of true wilderness. We reject the notion that wilderness is merely remote, scenic terrain suitable for backpacking. Rather, we see wilderness as the home for unfettered life, free from industrial human intervention.

Wilderness means:

- Extensive areas of native vegetation in various successional stages off-limits to human exploitation. We recognize that most of Earth has been colonized by humans only in the last several thousand years.

- Viable, self-reproducing, genetically diverse populations of all native plant and animal species, including large predators. Diversity at the genetic, species, ecosystem, and landscape levels is fundamental to the integrity of nature.

- Vast landscapes without roads, dams, motorized vehicles, powerlines, overflights, or other artifacts of civilization, where evolutionary and ecological processes that represent four billion years of Earth wisdom can continue. Such wilderness is absolutely essential to the comprehensive maintenance of biodiversity. It is not a solution to every ecological problem, but without it the planet will sink further into biological poverty.

The Wilderness Proposal: Core Reserves, Corridors, Buffers, and Restoration

We are committed to a proposal based on the requirements of all native species to flourish within the ebb and flow of ecological processes, rather than within the constraints of what industrial civilization is content to leave alone. Present reserves—Parks, Wildernesses, Refuges—exist as discrete islands of nature in a sea of human modified landscapes. Building upon those natural areas, we seek to develop a system of large, wild core reserves where biodiversity and ecological processes dominate.

Core reserves would be linked by biological corridors to allow for the natural dispersal of wide-ranging species, for genetic exchange between populations, and for migration of organisms in response to climate change.

Buffers would be established around core reserves and corridors to protect their integrity from disruptive human activities. Only human activity compatible with protection of the core reserves and corridors would be allowed. Buffers would also be managed to restore ecological health, extirpated species, and natural disturbance regimes. Intensive human activity associated with civilization—agriculture, industrial production, urban centers—could continue outside the buffers.

Implementation of such a system would take place over many decades. Existing natural areas should be protected immediately. Other areas, already degraded, will be identified and restoration undertaken.

The Wildlands Project sets a new agenda for the conservation movement. For the first time a proposal based on the needs of all life, rather than just human life, will be clearly enunciated. Both conservationists and those who would reduce nature to resources will have to confront the reality of what is required for a healthy, viable, and diverse North America. Citizens, activists, and policy makers will be able to confront the real choices because the choices will be on the agenda. It will no longer be possible to operate in a business-as-usual manner and ignore what is at stake.

The Wildlands Project will also inspire the development of indigenous proposals for other continents.

Who Are We and What Do We Do?

The Wildlands Project is a non-profit publicly supported organization with offices in McMinnville, Oregon, and Tucson, Arizona. We are a group of conservation biologists and biodiversity activists from across the continent.

We work in cooperation with independent grass-roots organizations throughout the continent, local chapters of national and international conservation organizations, and scientists and individuals to develop proposals for each bioregion. Development of regional Wilderness proposals is based upon principles of conservation biology. Draft proposals are developed through discussions and conferences that bring together regional activists, conservation biologists and other scientists, and conservation groups across the spectrum of the movement. The Wildlands Project supports this process through funding, networking, and offering technical expertise.

We undertake and encourage research on appropriate human activities in buffers, reintroduction of extirpated species, design of connecting corridors (especially through areas with significant human obstacles), overcoming fragmentation and achieving habitat connectivity, genetic diversity, population viability, and control of exotic species.

As proposals are developed we will publish the results in pamphlet form, in *Wild Earth*, and in other conservation publications to reach a wide audience. Videos, slide shows, and academic articles will be produced and traveling exhibits will be organized to educate the public about the proposals. When proposals for all bioregions of the continent have been completed, a book and compendium of maps will be produced, as well as updated videos and related materials.

In short, our job is to educate the public, the environmental movement, government agencies, the academic community, and others about the importance of biodiversity and what is necessary to protect it.

The Wildlands Project welcomes the participation and support of all persons and organizations interested in these issues.

Note

1. For more information write The Wildlands Project (a nonprofit organization) at 117 East Fifth Street, Suite F, P.O. Box 1276, McMinnville, OR 97128; (503) 434-9848.

The Wildlands Project:
Scientifically Sound,
Ethically Compelling—
and Politically Realistic

JAMIE SAYEN

With obligatory skepticism, David Quammen wrote of The Wildlands Project in the December 1993 *Outside*: "My first impression of this grand design has been that it's scientifically sound, ethically compelling, and politically impossible." How extraordinary that a plan to protect the ecological and evolutionary integrity of North America based on sound science and compelling ethics is unrealistic in our modern democracy. If attempting to live within the limits of physical and ecological reality is "unrealistic," what, one must ask, is "realistic"? Is the status quo either desirable or sustainable? Is it possible to continue living as we North Americans live today? I submit that the greatest obstacle to the realization of The Wildlands Project's plan is this obligatory skepticism of our friends who have swallowed the burden-of-proof scam that industrial society uses to delay or defeat anything that threatens the status quo. It is time we reverse the burden of proof.

Rather than dismiss The Wildlands Project's vision, let us scrutinize current "reality." Let the champions of industrial civilization demonstrate that destroying habitat, driving species and ecosystems to extinction, befouling soil, water, and air, and spewing toxic and radioactive wastes hither and yon is ecologically sustainable. Let them demonstrate that destroying indigenous cultures, foreclosing the options of future generations of all species, and condemning children to starvation, crack cocaine, network television, and the values of suburbia are ethical. I am always amazed that we allow the defenders of the indefensible to put us on the defensive. We who wish to conserve 3.5 billion years of evolving life are environmental extremists? We who would protect wild areas from corporate plunder are elitists? We who suggest that an economy based on environmental abuse is not sustainable are against jobs?

Please, David Quammen and all of you who are rightfully seduced by the breathtaking beauty and simplicity of The Wildlands Project's message—if an idea is scientifically sound and ethically compelling, then it must be realistic. If current political institutions, corporations, mainstream environmentalists, and other leaders are too corrupt, too immature, or too fainthearted to believe there is any room in our political calculus for physical reality, don't patronize the bold, the visionary, those who are modest enough to wish to share this planet with *all* our relatives from single-celled creatures to whales and sequoias. Instead of mocking us with obligatory skepticism, roll up your sleeves and help us change a "political reality" that is demonstrably crazy. For too long our culture has sought to bring ecological reality in line with our political and economic value system. It's time to try something new. Let's bring political reality into harmony with ecological reality.

Obligatory skepticism is a corruption of thoughtfulness; it is unthinking skepticism. As a participant in the development of The Wildlands Project, I can state with confidence that everyone associated with it welcomes thoughtful skepticism: Is this idea sound? Does the science bear us out? What are the ramifications of taking this action? Of not taking this action? The thrust of these questions is: how can we develop a continental strategy that effectively protects the natural diversity of life and its processes? Constructive skepticism is essential.

Obligatory skepticism is a symptom of our desire to be liked, to be respected, to be able to "retain a seat at the table" of power. It is a gross corruption of the political process to suggest that defenders of ecological integrity must be careful not to offend the sensibilities of the ecological ignoramuses who control modern industrial society. By allowing politicians and corporate executives to sit in judgment over our credibility, we concede to them the terms of the debate. And when we go up against *their* power on *their* terms, we lose.

An additional consequence of acquiescing to their standards of credibility is that the environmental community is divided and conquered. If some of us retain our credibility in the eyes of Speaker Foley and Senator Mitchell by agreeing to limit the terms of the debate to industry's preconditions, others of us who reject such self-censorship are dismissed as "politically unrealistic, environmental extremists." Without credibility, we are not permitted to play the game. The sellout of ancient forests in the Pacific Northwest continues. The betrayal of the Northern Rockies Ecosystem is assured. In the northern Appalachians, mainstream fears of challenging the validity of hydroelectric dams is dooming efforts to save the wild Atlantic salmon.

At a recent conference on "The Ecology of the Northern Forests," the discussion of landscape-scale strategies for protecting biodiversity provoked a graduate student to warn us that these were unproved theories, that scientists have an obligation to retain their objectivity, and that we should refrain from action until we can prove these theories. I exploded. Since when has your "objective" science shown a similar concern with the unproved theories that dioxin is harmless, that ancient forests are biological deserts, that hydroelectric power is clean? This poor victim of my wrath had fallen into the trap most environmentalists, scientists, and other citizens succumb to—the double standard that industry, polluters, and deforesters can do anything they want until *we* prove beyond a shadow of a doubt that irreversible harm has been done, whereas protection strategies cannot be implemented until we have proved them to be the absolute cure to the problem. This is objectionable, not objective, science.

———

For the past five years I have been absorbed in a fascinating drama revolving around the fate of the northern Appalachians and the Adirondack and Tug Hill regions of northern New York—the congressionally authorized Northern Forest Land Study (NFLS) and Northern Forest Lands Council (NFLC). From the outset, representatives of the timber industry from Maine, which contains roughly 60 percent of the four-state region, have threatened to pull out of the NFL process if certain issues were even discussed: clearcutting and forest practices, ecosystem health, wilderness. Other issues, such as new national parks and large-scale federal acquisition of millions of acres of paper industry lands currently for sale, have received biased treatment.

From the outset, a few of us showed large aerial photographs of industrial clearcuts, some the size of a township, and demanded that the NFLS and NFLC address the issue of forest practices. We were dismissed as naive environmental extremists. Because of industry, ecologists were invited to address the council only after four years had been squandered. Despite overwhelming evidence of abusive industrial forestry, forest practices must still be addressed through the council's back door. The council continues to resist honest discussion of federal acquisition—despite strong public support and even stronger scientific evidence of the need for large ecological reserves in a region that (except for Adirondack Park) has less than 7 percent of its land in public ownership and barely any in true wilderness. Instead, the council has devoted inordinate time and energy to studying ways to give more tax breaks to the timber industry and to "saving" timberlands via expensive conservation easements that often cost 50 to 90 percent of the full purchase price. These easements prevent construction of condos, but they permit the landowner to continue clearcutting, herbiciding, and high-grading.

Ironically, as the regional economy worsens, as more lands come up for sale, as the public grows more outraged by photos of township-sized clearcuts, the council finds itself on the defensive. Its credibility has been profoundly compromised, and council members know it. When confronted with a photo of a massive whole-tree harvest clearcut, the public rightfully asks: what sort of credibility can a Northern Forest Lands Council have if it has refused to study forest practices? Today, despite efforts by industry and "property rights" zealots to stifle free and open discussion, despite the council's lack of courage and vision, the public has forced the council to ask (albeit timidly): how can we establish a network of connected, buffered ecological preserves that assure the ecological and evolutionary integrity of the region in perpetuity?

Reserves in the Northern Forest region must encompass 10 to 15 million acres of land in the 26-million-acre region. Currently 2.4 million acres of "forever wild" state forests are publicly owned in the Adirondacks. There are only another 1.5 million acres of multiple-use public lands (including some, but not much, wilderness) in the rest of the region. How do we add another 8 to 10 million acres to the public land base for wilderness? Currently 3 to 5 million acres of industry lands are for sale or considered "nonstrategic" and potentially available for sale at an average price of $200 to $250 an acre. As the paper industry redirects more and more of its investment to the southeastern United States, hundreds of thousands, probably millions, of additional acres will be offered for sale by Champion, International Paper, James River, and others.

Recently a colleague and I proposed the establishment of a 5-million-acre (or greater) Thoreau Regional Wilderness Reserve in the uninhabited northern reaches of Maine that includes the St. John and Allagash river watersheds, the Greater Baxter State Park ecoregion, the East and West Branch Penobscot River, as well as connection links to other reserves in Maine, New England, eastern Canada, and the Gulf of Maine. (See *The Northern Forest Forum*, Spring Equinox 1994.) Although the mainstream environmental groups of the region are not yet ready to support it publicly, many mainstreamers privately saluted us for the proposal. A few asked why our proposal was so modest.

There is growing public support for wilderness, wolves, and reserves. When the public discovers that 10 million acres of industrial forestland can be purchased for the price of about two and a half Stealth bombers, it will recognize a genuine bargain. In 1992, with hardly any public debate, Congress forked over $7 billion to rebuild southern Florida after Hurricane Andrew so that another hurricane can come along and inflict $8 billion in damage. For one-third the cost of temporary relief from Andrew, we could establish vast ecological reserves in perpetuity.

Yesterday this idea was so far-fetched that no self-respecting environmental group would even consider it. Today the local, regional, and national environmental groups working on Northern Forest issues agree that a large network of ecological reserves is necessary. Not all of them are ready to embrace the scale of The Wildlands Project's proposal, but many do so in private, and all agree that we need many big reserves throughout the region. Moreover, they have discovered that "wildlands" has a sweeter ring to it than "priority conservation areas." What was utterly impossible a few years ago will come to pass in our lifetime if we stop the self-censorship of legitimate dreams.

The moral of the story for grassroots environmentalists is this: do your homework scrupulously, relentlessly, and imaginatively. Persist. Show up at every possible occasion. Put your trust in the truth—both ecological and ethical. Draw strength from the knowledge that the general public will gradually listen to your respectfully advanced message; it will be grateful to you for filling a void with a wildlands-scale vision. At length, the good old boy network will have to deal with our agenda. The secret to persistence is a sense of humor—easy to maintain when listening to the nonsense of those who are paid a lot of money to defend the indefensible.

———

I have been discussing examples of how the power elite subverts democracy by keeping legitimate issues off the discussion table. The crudest way is through bullying and threats of violence. This is the preferred method of the "property rights" or "wise use" movement. In our region, these bullies have shut down meetings with threats of violence, slashed tires, and resorted to John Birch Society-style red-baiting. You may have noticed that the "international communists conspiracy" has been replaced in the 1990s by the "international environmentalist conspiracy" in the literature of paranoia.

Economic blackmail is slightly more subtle, much more socially palatable, and a far more effective means of subverting democracy and protecting illegitimate corporate power. The argument is simple: any regulation over clearcuts or dioxin discharges into rivers will cost jobs. Such threats are designed to stifle any further discussion. For the past century this strategy has succeeded, and, as a result, society has never seriously examined the validity, or ethics, of such a stance. These morally reprehensible threats present us with a false dichotomy: jobs or environment. Apparently, industrial civilization believes there are no safe jobs on a healthy planet.

What is so silly about these threats is their irrelevancy. After the passage of the Clean Water Act in the early 1970s, the paper industry in Maine did not go out of business as its paid lawyers and liars had threatened ad nauseam—indeed, the industry expanded operations. In the 1980s jobs did disappear, even though there were no significant new environmental regulations in Maine and the economy was booming (thanks to the Reagan era for plundering the options and rights of future generations of all species). Why did jobs disappear during boom times? Automation in the mills and mechanization in the woods—and global competition, the preferred euphemism of environmental plunder in the 1990s.

Future generations will look back at our pathetic obsession with growth-at-any-cost economics and wonder how we could have invented such a disingenuous accounting system. Polluting is profitable only if you can pass the costs onto someone else. If the paper industry had to pay the cost to ecosystems, human health, and foreclosed economic options from dumping dioxins and hundreds of other organochlorines into the rivers, it would convert today to a chlorine-free bleaching process or would abandon altogether the conceit of bleaching paper. Do you really need lily-white toilet paper? David Brower has suggested that if the Bureau of Reclamation wants to dam the Grand Canyon, it should first build an exact replica of the Grand Canyon, down to the last soil microbe, before it begins to impound the Colorado. How cost-effective would that "clean" hydropower be?

On the local level, if you believe the pundits in New England every March—Town Meeting Time—democracy is alive and flourishing. In fact, Jefferson and Lincoln would be unable to find its pulse today. Town Meeting is the scale on which democracy should work well, but democracy depends upon free and open discussion, by an informed electorate, of issues it understands and can exercise local and regional control over. In reality, Town Meeting is a mockery of democracy. My vote counts for little if I don't understand the issue, or if I have been presented with false choices by the state legislature and the absentee corporations that own most of northern New England and control the economy. If our selectmen are ill informed or have conflicts of interest, and if free and open discussion is trampled by residents who promote obdurate Yankee fiscal policies, property rights scare tactics, and the like, the exercise is facetious.

North American democracy is at a crossroads. Just as the Civil War was an inevitable resolution to the unfinished business of establishing an independent nation that countenanced slavery, there must be a reckoning for a society whose economic, social, and political institutions are based on ecological abuse. When Lincoln delivered the Gettysburg Address, he challenged the living to dedicate themselves to the "unfinished work" of actualizing the intent of the Founding Fathers in 1776—that the experiment in self-government by free men and women who were created equal should survive and flourish. Today we are in the midst of another great war: a war against 3.5 billion years of evolving life. Just as Lincoln's generation was called upon to acknowledge the need for evolving new values and institutions to realize the work of the Founding Fathers, our generation must measure up to a political *and* biological challenge.

The paradox of freedom is that we possess the political freedom to make these needed changes peacefully, equitably, but we do not have the biological freedom to ignore, or to cut a compromise with, the ecological crisis that now grips us. To salvage democracy, our generation must lead the third great democratic revolution in our history. We must bring forth a biological democracy that assures equal standing for all living beings and unborn generations of all species.

The Wildlands Project meets this challenge with generosity, humility, and the spirit of playfulness so absent in the political machinations of the environmental bean counters and slaves of credibility. Its first tenet is: Assure the dance of evolving life. All other matters, economic, social, political, cultural, must develop within the limits of physical and ecological reality, not in denial of these limits. The great irony is that by subsuming anthropocentric ego and drives, we, at long last, have hit upon the survival strategy that optimizes chances for human happiness—whereas the allegedly humanistic and anthropocentric values of current culture have wrought unparalleled human misery in this century of genocide, famine, and Superfund sites.

———

David Quammen concluded his article on The Wildlands Project with the warning that although the project has "great merit, . . . it will never succeed and probably shouldn't, I suspect, if the efforts to realize it becomes another us-versus-them political battle, since us-versus-them is just another form of fragmentation. It can only succeed through political persuasion." Good advice. But again I fear that Quammen allows the ecological abusers and their codes of political credibility to set the terms of the debate. It's the old "blame the messenger" story. If The Wildlands Project polarizes the debate because its sound scientific prescription for self-preservation provokes a reaction from plunderers who will be inconvenienced by biological democracy, should the project keep quiet?

Let's face it: The Wildlands Project is going to stir up a storm because industrial civilization is incompatible with biological diversity. Quammen, however, is correct in his admonition to avoid "avoidable" controversy. Let us heed the advice of Sun Tzu, author of *The Art of War*, who said twenty-five hundred years ago: "Therefore those who win every battle are not really skillful and those who render others' armies helpless without fighting are the best of all." We can win support for our ideas without debilitating battle, posturing, and polarizing. To do so we must address the concerns and fears of our neighbors, we must patiently clarify misunderstandings, and we must welcome constructive criticism and suggestions that strengthen the proposal or rescue it from error. The Wildlands Project must have regional flexibility so that concerns unique to New England or the Great Plains can be addressed.

We must set the terms of the debate. For too long, David Brower notes, environmentalists have sought merely to slow the rate of environmental degradation. We must turn things around; we must accelerate the rate of healing. What I have found most striking in the Northern Forestlands debate is the absence of vision. The timber industry knows what it wants and knows how to get its tax breaks, subsidies, and exemptions from paying the costs of "environmental externalities." The environmentalists have been so caught up in the trap of trying to slow the rate of degradation (and protecting their credibility) that they have thus far failed to come up with an overarching vision that liberates us all from the ecological dead end that civilization has become. The politicians are elected by corporate money, pander to right-wing zealots, and have insulated themselves in this cocoon of "political realism" that is utterly detached from physical reality and has become the antithesis of democracy.

Into this void we offer the vision and plan of action of The Wildlands Project. Breathtaking in its audacious simplicity, it already has earned a substantial following because it is scientifically sound and ethically compelling and because no other realistic vision has been offered. By default, The Wildlands Project is setting the terms of the debate. Some love it; some hate it; but all are discussing it. This reminds me of the politician who said, "I don't care what you print about me as long as you spell my name correctly." At this stage it is more important that people discuss the project than that they embrace it. As the idea becomes more familiar, less implausible, less threatening, the support will swell. For now, debate and discussion keep the idea alive. Some critics will oppose anything we say; thoughtful critics will make great contributions to our evolving vision—but only if we continue to offer a vision grounded in ecological reality, not one truncated by fatuous worries about political credibility or fears of the attacks from the "wise-use" (wild abuse) movement.

I do not believe that "patient persuasion" will succeed, however, nor that it is necessary. I can't persuade A to fall in love with B, and it is foolish to try. More to the point, why should I waste your time and mine trying to convince you to assimilate something already at the core of your very being? A line from a Talking Heads song says, "Underneath the concrete, the dream is still alive." Within your heart and mine, wildness survives. It may have been polluted and deformed by our consumer culture, but it hasn't been eradicated. There is no need for me to persuade you to recreate that wildness within. But those who believe in The Wildlands Project can serve as catalysts for people rediscovering the wildness within.

We ran a gorgeous photo of a wolf on the back cover of the second issue of *The Northern Forest Forum*. When I went down to the ancient web press that prints the *Forum* to pick up that issue, one of the pressmen, a hunter living in a conservative rural New Hampshire community, approached me and, pointing to the wolf, said: "This is what I like about your paper." "I'd prefer the real thing to a picture," I responded. A big smile broke over his face, and he gave me a thumbs up sign. It had never occurred to him that it was possible to dream of the return of wolves to northern New Hampshire after more than a century's absence. I had persuaded him of nothing. The photo of the wolf had reconnected him with something inside that he loved—a generous and wild spirit. Underneath the breastplate the dream is still alive.

Many people who have made this reconnection with the wild have expressed gratitude to those of us who dream aloud. I believe there is a powerful psychological force at work here—namely, to support the return of wolf, wolverine, and wilderness is an act of generosity. Yes, we are saying, there is room in this region for cougars, lynx, and old growth; we humans *can* share. Reconnecting with the wild liberates one from the selfish, economic-centered values of our culture, and this affirming experience is building a powerful political, cultural, and moral movement.

A movement built on generosity of spirit, faith in ecological reality, and reconnection with the wild within and without will endure long after the "property rights" fearmongers and truth abusers have disappeared into the compost pile of history. In the short run it is easier to stir up negative sentiments through fear tactics than it is to stimulate support for visionary, life-affirming values. But it is hard to sustain anger and lies. The truth endures—and supporters of an ethically compelling vision grow stronger with the passage of time. One lesson we should learn from the misguided anger of the "property rights" (wild abuse) crowd is: don't get mad, get effective.

If Quammen is wrong about "patient persuasion," he is right to insist that advocates of The Wilderness Project must persist in addressing a whole range of issues—from wilderness protection and restoration and economic sustainability to the establishment of biological democracy. Critical to the project's success will be our ability to envision an economy that is circumscribed by the limits of physical and ecological reality. So long as transnational corporations rule our various ecoregions, the economy will continue its war against the wild. But it also wars against my rural neighbors whose present and future options have been sacrificed to the interests of these corporations. So long as the absentee owners control our region's destiny, our cultural crisis—which is an absence of an earth-based, genuinely regional culture—will worsen. Lack of control leads to cultural despair, cultural despair keeps us divided: politically impotent colonists of absentee plunderers.

In Coos County in northern New Hampshire, where I live, we have no school of higher education, no school of natural or cultural history, no interactive museums, scant support for artists, musicians, dancers, and storytellers who would promote a sense of place, a sense of pride in place. Instead we have the culture of the satellite dish, the snowmobile, and the beer can by the roadside. Currently we have near total "brain drain" in this region. A while back I asked the local high school principal how many of his graduates who went to four-year programs later returned home to live and work. His bone-chilling answer: "approximately zero."

The Northern Forest Forum has proposed a strategy for cultural restoration (Mud Season 1993, vol. 1, no. 4) that would preserve, protect, and promote a flourishing earth-based culture in the homes, in the schools, and in the products of local community-based agricultural endeavors, woods workers, and other craftspeople. Rather than promote another community college whose goal is to produce more accountants, insurance salesmen, and real estate agents, we have proposed the establishment of a Coos County Ecological Restoration Academy to offer programs in regional natural and cultural history and restoration. Additional programs in ecologically sustainable forestry and business should replace the current batch of growth-at-any-cost business degrees currently offered by nearly every college and university in the country.

Aligned with this Restoration Academy should be an interactive natural history museum. Smaller, satellite museums could be scattered throughout the region. North Stratford, New Hampshire, could house a Log Drive Museum. Island Pond, Vermont, could establish a railroad museum in its fine old railroad station. Museums celebrating cultural diversity, local agriculture, and regional history could help restore regional identity and pride and serve as a far more attractive destination for tourists than Santa's Village and Six Gun City currently offer.

These cultural, economic, and political reforms are vital if we are to realize The Wildlands Project's vision before much more irreversible ecological damage is inflicted. They are the human context of the vision, and the project must never lose sight of their import, although such reforms will take time and will require resources not readily available. The primary job of The Wildlands Project is, and must remain, wilderness restoration and protection. That is our area of expertise.

We will need help from watershed councils, from gifted writers such as David Quammen, and from mainstream environmental groups. In the context of the economic growth agenda of the Reagan-Bush-Clinton era, The Wildlands Project is, admittedly, audacious. As currently formulated it has many unresolved problems and issues. But if the science upon which it is based is sound and the ethics of the vision are compelling, then men, women, and children of goodwill and generous spirit will find ways to work together to test and strengthen it.

The Wildlands Project needs thoughtful critics. Thoughtless criticism designed to protect the critic's credibility in the eyes of the power elite will delay, perhaps derail, realization of the dream. As you judge the aims of The Wildlands Project against the status quo, do not confuse political and ecological realities. Do not attempt to impose a political compromise upon an ecological issue. We can repeal a tax law, but not the law of gravity. The timber companies, Congress, and the U.S. Forest Service can decree that we will cut so many billions of board feet per year. But they cannot command the impoverished, eroded soils to grow that amount of wood. Until our political and economic institutions recognize the necessity of living safely within the limits of regional and planetary carrying capacity, they will remain scientifically unsound, ethically repellent, and politically impossible.

The beauty of wilderness restoration is that, for the most part, it will follow naturally, if we allow it. Long ago, Lao Tzu said:

> The sanest man . . .
> Takes everything that happens as it comes,
> As something to animate, not to appropriate,
> To earn, not to own . . .

Population Growth and the Wildlands Vision

MONIQUE A. MILLER

Rapid human population growth in the United States could turn The Wildlands Project's vision into a pipe dream, and soon. Yet considering the ecological importance of restoring wilderness on a continental scale, how can we work together to address human population growth as one of the foremost challenges to the viability of the Project?

It is certainly true, as David Clarke Burks observes, "that an over-emphasis on human concerns, at the expense of other life forms, will ultimately backfire because human well-being is tied to the well-being of all life on earth."[1] But as a practical matter, it is equally true that the ultimate success of The Wildlands Project will be entirely decided by the will and number of *Homo sapiens*. It will be difficult enough to get a sufficient following of fellow human beings to support the wildlands idea. Unless concrete steps are taken now to reverse current U.S. population growth trends, not only will our present opportunity to realize the wildlands vision be compromised, but our ability to preserve and protect wildlands in the future will be dramatically reduced or even rendered impossible.

Let us look at the fundamental reality: the United States is the fastest-growing industrialized nation in the world. We are presently adding to our country approximately 58,000 people every week—a total of 3 million people every year. In other words, we are adding the population equivalent to the size of Washington, D.C., to our country every three months. The natural increase of the U.S. population (births minus deaths) is now approximately four people per minute. Adding an estimated net immigration of two people per minute, this means the United States is growing by approximately six people every sixty seconds.[2] We already rank third in the world only after China and India in terms of the size of our country's population.[3] Indeed, the U.S. Census Bureau projects that the U.S. population may more than double in number to one-half billion people within the next sixty years.[4]

The problem is that our current population growth rate may actually accelerate. Although the U.S. fertility rate is now averaging two children per woman, or "replacement level,"[5] the absolute number of births in 1990 was 4.18 million,[6] a figure disturbingly similar to the peak of the Baby Boom, which marked the highest number of births ever recorded in U.S. history.[7] Moreover, the world's population is now growing by approximately 95 million people per year—at precisely the same time that an increasing number of nations around the world are closing their borders to immigration.[8] Although the United States already accepts more immigrants than any nation in the world,[9] one could argue that pressure to accept even more in the future is likely to increase.

What does population growth have to do with protection of biodiversity? Almost everything. Although there is a woeful lack of research proving that human population growth is the ultimate cause of our plummeting loss of biodiversity in this country, logic and evidence would suggest that there is an undeniable correlation between the two. Consider the following examples.

In the past fifty years—less than a single human lifetime—Florida's population has increased more than sixfold.[10] Almost 17 percent (111 species) of all native fish, reptiles, amphibians, birds, mammals, and invertebrates are now thought to be in danger of extinction in the state. Nearly half of all Florida's nonmarine vertebrates are known or suspected to be declining. Today the state only has thirty to fifty Florida panthers and five hundred to a thousand black bears left.[11] What is the chance of any threatened species surviving in a state that has recently been adding an average of 950 new residents every day,[12] especially if those threatened species are Florida panthers, which are said to require

50,000 acres, or black bears, who need 15,000 acres, just to survive?[13] Florida's population is now growing on average by 2.8 percent annually. If that rate of growth continues, Florida's population is projected to double in size within twenty-six years.[14]

In California, the population has grown by 30 million people since the 1880s. Since then it is estimated that thirty-seven species have become extinct in the state.[15] (One of the few mammals that still outnumbers California's human population is the prolific wild rabbit.)[16] Although the grizzly bear is still proudly displayed at the center of California's state flag, not one wild grizzly can be found anywhere in the entire state. At the present time, over 100 species are listed in the state as threatened or endangered.[17] California's human population is now growing on average by 2.1 percent annually. If that rate of growth continues, California's population is projected to double in size within thirty-three years.[18]

Hawaii, to take a final example, is becoming known as "the extinction capital of the modern world." Seventy of the 140 species of native birds have disappeared since humans arrived on the islands.[19] Hawaii's human population is now growing on average by 1.4 percent annually. If that rate of growth continues, Hawaii's population is projected to double in size within the next fifty years.[20]

Looking back through American history, we find the United States has lost half its wetlands, 90 percent of its northwestern old-growth forests, and 99 percent of its tall grass prairie within the past two centuries. Possibly as many as 490 species of native plants and animals have vanished as a result, and another 9,000 species are now at risk.[21] Although poor land use decisions also contributed to this massive habitat loss, there is a more obvious cause: the U.S. population since 1790 increased from under 4 million to nearly 260 million people today. As a more recent example of ecological destruction, the first U.S. endangered species list compiled in 1967 had sixty-seven entries. As of 1992, seven hundred were listed, slightly more than half of them plants and invertebrates.[22] During that time the U.S. population grew by more than 55 million people.[23] Across the country now, about three thousand plants (nearly one in every eight native species) are considered to be in danger of extinction. More than seven hundred are likely to disappear in the next ten years.[24] Assuming that population continues to grow only at current rates, the number of people is projected to grow by more than 110 million within the next fifty years.[25]

Projections are not predictions, of course. But without quick action to reverse current trends, the implications of continued human population growth for biodiversity would seem to be ominously obvious.

A few scientists do recognize and sometimes directly address the human connection to habitat loss. One of the more recent examples is C. Dale Becker's comment: "During the last 100 years, 40 known taxa of North American fish have become extinct from activities related to occupation and development of habitat by humans, and their extinction rates are expected to increase."[26] Similarly, John Terborgh of Duke University notes: "Increasingly, research is indicating that human activities have directly and indirectly altered the environment in ways that are detrimental to birds."[27] Although we can be encouraged that the human connection to biotic impoverishment is made, it must still be acknowledged that neither example directly mentions human population growth as a contributing, much less a primary, cause.

How are we to demonstrate the link between population growth and rapid loss of biodiversity in the United States without cross-disciplinary and potentially controversial scientific research? Gary Meffe of the University of Georgia, Anne Ehrlich of Stanford University, and David Ehrenfeld of *Conservation Biology* muse: "Since the first issue of *Conservation Biology* was published in May 1987, . . . 177 contributed papers, 22 essays, 27 notes, and numerous comments, Diversity columns, editorials, and other articles [were printed]. Yet not one of these contributions directly addressed human population growth, a root cause of our collective ecological and social disasters. This omission is not because of editorial policy; no papers on human population growth have been received."[28]

One can see why conservation biologists have limited themselves to descriptions of habitat conversions (such as wetland dredging and filling, urbanization, agriculture, silviculture, or mining) as the explanation for our rapid loss of biodiversity. Such statistics are easier to collect, make a superficially more direct correlation between cause and effect, and are certainly less "political" than attempting to demonstrate that increasing habitat loss in the United States is caused at least partially by our own species' dramatic population increase. But conservation biologists must begin to write about the population growth/diversity loss connection and continually search for the "chain of causation" in habitat destruction.[29] Without progress in this field, statistics would suggest that all other work by conservation biologists may ultimately be for naught.

In addition to such documented field research, the environmental community must recognize that each activist should work to ensure population stabilization and eventual reduction as quickly and humanely as possible. In other words, all environmentalists should become population activists—some of them just do not know it yet. For those who have already made this connection, what options remain for supporters of The Wildlands Project who want to work *now* to mitigate the effects of population-related habitat loss?

———

Some wilderness advocates believe that the establishment of "boundary lines around . . . our ever-expanding concentrations of civilization and industry" could be a step in that direction.[30] It is estimated that at least five and a half square miles of rural land is converted to urban, suburban, and other forms of development each day. Annually this amounts to over 1.2 million acres lost, an area larger than Grand Canyon National Park.[31] Would high-density living help reduce such habitat loss?

Much has been written about the immediate resource efficiencies associated with high-density living, and it is true that the high-density option is ecologically more benign in some respects. But simply limiting human transformation of the landscape to within a certain growth boundary fails to acknowledge that the key criterion is not land area but *carrying capacity*. The carrying capacity concept explains that there is an optimal and also an ultimate limit to the number of people who can be supported in a given land area without degrading the environment. Using this concept as a guiding principle, it becomes obvious that no matter how tightly packed within, or sprawled outside, an urban area people choose to live, any city which is not self-sustaining can survive only by the constant importation of resources and exportation of wastes. This two-way transfer—the importation and exportation of carrying capacity—requires vast areas of land outside the strict urban boundaries and also simultaneously reduces the carrying capacity of those external sources and sinks as well.

It is understandable that those who wish to see the wildlands vision realized might support the idea of high-density living because they want to believe it would destroy less wildlife habitat than any other human settlement option. But concentrated human development cannot compensate for the fact that an ever-increasing number of people ultimately requires an ever-increasing amount of resources, pollution absorption capacity, and, ultimately, an ever-increasing amount of arable land. There is an environmental impact to almost every human action—and, therefore, to every additional human being. Any environmental savings we might be able to achieve as the result of high-density living will ultimately be negated if human population continues to go unchecked in this country. Therefore, any attempt to encourage high-density human settlement in order to protect wilderness or farmland is ultimately futile unless we first stabilize the human population.

———

Might there be other solutions to the pressing problem of human population growth and loss of biodiversity in the nation? Although the idea of somehow reversing the population growth curve can be initially daunting, the fact remains that immigration accounts for approximately 50 percent of this country's population growth.[32] Therefore, one of our options to reverse current population growth trends, albeit a controversial one, would be to restrict immigration into the United States. This suggestion should not be misconstrued to mean that the country should close the border to all immigration. Instead, it is intended to encourage public debate on the issue of what the optimum population of the United States should be.[33]

One could of course argue, with Paul Ehrlich of Stanford University, that we should work to reduce U.S. fertility rates before reducing immigration into this country. This choice is wrong-headed for two reasons. First: If we in the United States were to insist that immigration levels be brought down to a more sustainable "replacement level," [34] we would definitely see the eventual stabilization of our human population because we have already achieved replacement level fertility (approximately two children per woman) in this country. Second: Reducing the fertility rate below this current average of two children per woman may well be possible over the long term. [35] But getting enough Americans to voluntarily agree to have on average fewer than two children—in time—could be an insurmountable task whereas it would require only one act of Congress to restrict immigration and thus cut the U.S. population growth rate in half.

Wildlands advocates who consider themselves political liberals (and therefore viscerally reject the idea of immigration limitations) must accept the fact that they are also ecological conservatives. This realization demands an answer to one very important question: In a country of increasingly limited resources, how can we possibly provide for the needs of America's nonhuman species if the human population of this country continues to grow, regardless of whether that growth is the result of native births or immigration?

A second solution could perhaps answer a very legitimate disagreement with The Wildlands Project as currently proposed. As suggested in the magazine *Science*, this "controversial plan to protect North American biodiversity calls for nothing less than resetting the entire continent. That may be too much to ask of the people who already live there." As the authors point out, "the Wildlands approach calls for 23.4% of the U.S. to be returned to wilderness, and another 26.2% to be severely restricted in terms of human use. Most roads would be closed; some would be ripped out of the landscape. The plan does not specify what would happen to the nearby inhabitants." [36]

If it is "wildly impractical" [37] for us to consider relocating millions of human beings, is there another way to achieve what is ecologically vital and have it still be practically feasible? The Buffalo Commons proposal of Frank and Deborah Popper of Rutgers University makes sense. Based on an initial observation of changing demographic and economic trends, the Poppers have developed in effect an expansive wilderness recovery strategy just by documenting those areas which are in the middle of a natural depopulation cycle. After collecting county data on short-term and long-term population loss, low population density, aging, poverty, and per capita investment rates and charting the results of this documented "land-use distress," the Poppers found one region of the United States where 110 counties show at least three of these six signs overlapping: the Great Plains. Stretching from most of eastern Montana and western North Dakota all the way down south through Texas, the Poppers' research shows that the counties of greatest land-use distress tend to cluster in the Dakotas, Montana, Nebraska, and Texas.

It must be noted that these natural depopulation cycles the Poppers have documented are largely the result of many people choosing or being economically forced to forgo a rural, agricultural existence in favor of an urban lifestyle. Therefore, the regional decline in the rural population has not reduced the aggregate size of the human population at all. (See my earlier comments on the high-density living option.) Nevertheless, the most important lesson we should take from the Buffalo Commons idea is the vision of how much easier our work to establish a network of wilderness reserves and corridors would be if only human population size in the United States were slowly decreasing instead of moving rapidly in the opposite direction.

Considering that human population growth has had such a consistently devastating impact on biodiversity and that human population in this country is continuing to grow more rapidly than ever before, time for massive restoration of wilderness is quickly running out. Wilderness advocates must work together to create concrete strategies that will assure eventual U.S. population stabilization and then reduction. This means unequivocally that wildlands advocates are going to have to become involved in one of the most challenging and controversial issues that has ever confronted the human species. But unless steps are taken now to reverse current population growth trends, it is difficult to imagine that even small components of the wildlands vision, much less the ultimate objectives of the project, will ever be realized.

Notes

1. David Clarke Burks, personal correspondence, 19 August 1993.
2. Population Reference Bureau, "Population Data Sheet," Dec. 1992.
3. Population Reference Bureau's 1993 World Population Data Sheet.
4. "Population Projections of the United States, by Age, Sex, Race, and Hispanic Origin: 1992 to 2050," Report:25–1092.
5. Replacement level fertility is defined as 2:1. Allowing for infant mortality, this works out to be approximately two children per woman.
6. National Center for Health Statistics *Digest* 40(8) (12 December 1991).
7. The peak period of the Baby Boom fell between 1959 and 1961 according to the National Center for Health Statistics, *Advance Report of Final Natality Statistics*, 1989, "Births, Marriages, Divorces, and Deaths for 1991," vol. 40, no. 12.
8. Rosemary Jenks, ed., *Immigration and Nationality Policies of Leading Migration Nations* (Washington, D.C.: Center for Immigration Studies, 1990).
9. Center for Immigration Studies. The Immigration Law of 1990 increased immigration to the United States by 40 percent annually. According to the Federation for American Immigration Reform, this level is higher than ever before in U.S. history.
10. According to the U.S. Census Bureau's Population Division, Florida's population in 1940 was 1,897,414; by 1990 it had grown to 12,937,926.
11. "Conserving Florida's Biological Diversity: A Report to Governor Lawton Chiles," by the Florida Biodiversity Task Force, February 1993.
12. Calculated from Population Reference Bureau's "United States Population Data Sheet," Dec. 1992 based on Florida's resident population on 1 July 1991 of 13,277,000 and its resident population on 1 April 1990 of 12,938,000.
13. "Wildlife Depends on River Corridors," *Coastal Heritage* (Summer 1992):10.
14. Calculated from Population Reference Bureau's "United States Population Data Sheet," December 1992, based on an average annual percentage change in population between 1 April 1980 and 1 April 1990.
15. California Department of Fish and Game as collected by Californians for Population Stabilization.
16. Ibid.
17. Ibid.
18. Calculated from Population Reference Bureau's "United States Population Data Sheet," December 1992, based on an average annual percentage change in population between 1 April 1980 and 1 April 1990.

19. *Nature Conservancy* (March/April 1992):28, taken from U.S. Fish and Wildlife Service and Hawaii's Department of Land and Natural Resources.

20. Calculated from Population Reference Bureau's "U.S. Population Data Sheet," December 1992, based on an average annual percentage change in population from 1 April 1980 to 1 April 1990.

21. *Environmental Almanac*, World Resources Institute, p. 159.

22. Personal phone call to Division of Endangered Species, U.S. Fish and Wildlife Service.

23. Calculated from phone call to Population Division, U.S. Census Bureau, based on a 1967 population of 198,428,000 and a 1992 population of 256,337,000.

24. WorldWatch Institute press release, 25 April 1992.

25. Calculated from figures given by the U.S. Census Bureau regarding the 1993 population of 257,592,000 and the projected population at current growth rates in 2043 of 369,857,000.

26. C. Dale Becker, "Population Growth Versus Fisheries Resources," *Fisheries* 17(15) (Sept./Oct. 1992):4.

27. John Terborgh, "Why Songbirds Are Vanishing," *Scientific American* (May 1993):98–104.

28. Gary K. Meffe, Anne H. Ehrlich, and David Ehrenfeld, "Human Population Control: The Missing Agenda," *Conservation Biology* 7(1) (March 1993):161–169.

29. Phrase suggested by Jim Fandrem, personal telephone conversation, 10 December 1993.

30. Les U. Knight, "Immigration: Where Do We Draw the Line?" *Wild Earth* 3(1) (Spring 1993):11.

31. *Developments*, Summer 1992, National Growth Management Leadership Project.

32. Population–Environment Balance, Washington, D.C. This figure also takes into account fertility rates of immigrants, which tend to be higher than those of native-born Americans.

33. Dave Foreman believes that an optimum population for humans would exist if all native large predators were present in healthy, linked populations. A country or region is overpopulated, in his opinion, if cats, bears, wolves, crocodiles, and such have been extirpated, are in decline or in small fragmented populations, or are threatened with such; *Wild Earth* 3(1) (Spring 1993):1.

34. Immigration is considered to have reached "replacement level" once the number of Americans leaving the United States to live in other countries every year is equal to the number of foreigners who move to the United

States annually. "Replacement level" immigration is estimated to be approximately 200,000 people.

35. Such a shift might come about either as the result of education or because of a cultural shift toward lower-than-replacement fertility (as demographers have documented in many other industrialized nations).

36. Charles C. Mann and Mark L. Plummer, "The High Cost of Biodiversity," *Science* 260 (25 June 1993): 1869–1871.

37. Ibid.

Humans and
Big Wilderness

MITCH FRIEDMAN

Siberian tigers are inclined to scent trees which angle above their trail. I sniffed the fresh urine on this oak, removed a few orangish hairs from the bark, and allowed tiger images to course a tremor through my veins. Tigers need far more habitat than the Far Eastern leopards whose tracks we were also noting. Knowing that tiger passed through here affirmed the wild stature of Khadrovia Pad, a nature reserve linked to Vladivostok by a few hours' rail passage.

I spent a whirlwind month in Siberia and the Russian Far East, an ecologist's working vacation, traveling with American friends who work their helping grassroots environmentalists organize. My purpose was to experience the famed massive tracts of unspoiled wilderness—real wilderness, *big* wilderness, full of brown bear, sable, tiger—that make the word Siberia itself seem larger than life.

I am a believer in wilderness—areas where evolution and ecological processes continue unfettered by the exploitation of industrial humanity. I have worked hard and long to protect such areas at home in the Northwest not only from those who move to cash in on nature's stock but also from the techno-arrogant who promote management (siviculture, wildlife management) as a misguided form of philanthropy for down-and-out ecosystems. Siberia offered the prospect of finally finding nature beyond human reach.

But this excursion tore deep gouges in my notions of big wilderness and human society. Unsettled musings continue to mold my impressions, like scar tissue lining those open wounds. Tugging at either end are conflicting conclusions: Are Russia's wildlands more damned than those of North America, or is the reverse true? America is the more fortunate society of the former two superpowers. My appreciation for democracy and dissent has swelled, but so too has my puzzlement about how our use of dissent has failed to provide security for the natural world.

ANOTHER WORLD

Khadrovia Pad will be a casualty of the Russian revolution, whatever its eventual outcome. During the Soviet years, human access was restricted. Now a failed economy and collapse of the police state undermine any law. The reserve is less than 20,000 hectares and surrounded by small communities. It is also walking distance from the Chinese and North Korean borders. Tigers and leopards, like too many of earth's species, have parts which some believe help men "get it up." Asians pay a premium for this seemingly common pleasure, while a tiger hide brings $15,000 (enough to support a Russian family for two decades) on the black market. An unstable society can quickly run through the last thirty to fifty leopards and two hundred or so tigers.

Russia also has massive unroaded regions—residual benefits of a centralized, unmotivated, and inefficient nation which can barely maintain infrastructure within its industrial cities. The Soviet Union left a massive legacy of deadly environmental problems: nuclear and toxic pollution of land, air, and water. But that state could not have consumed the endless forests and wilderness of Siberia had it aimed to. In any case, the Soviet era is history.

Russia's political structure may not stabilize within our lifetimes. The history, institutions, even philosophies, are not in place for basic democracy. Many older people even yearn for the good old days under Stalin. Things did not work very well before; now they don't work at all. Activists have incredible obstacles to overcome, but they nonetheless manage to sustain optimism and dignity in the face of despair. Today's unstable Russian society can be brutish, opportunistic, and violent. Mafia-like former KGB musclemen operate unchallenged. Corrupt monopolies formed of privatized former bureaucracies and cooperatives horde and consume vast resources. Foreign corporations, poised like jackals, wait for an open shot at the fresh kill. But even the likes of Weyerhaeuser cannot break in because the residual Russian timber industry (distinctions are obscure between the government agency and logging ventures) does not want foreign competition.

While domestic logging interests move without permits or enforced regulations into wild country, abusing land and labor shamelessly, they remain hampered by fuel shortages, outmoded equipment, and a lethargic workforce prone to take refuge in vodka. I suppose these parasites can do a lot of damage even without efficient equipment and foreign capital, but I doubt they can destroy the big wilderness that still exists. Transnationals like Mitsubishi and Hyundai, on the other hand, could make fast work of Siberia's oil, gas, minerals, and forests. But they may never get a stable opportunity to punch in the roads and build the facilities needed for competitive efficiency. Today's global economy means targeting wherever and whatever resources remain abundant, so modern carpetbaggers will circle for a long time before giving up in frustration. The future is anything but clear.

HOME LESSONS

We hired a helicopter to fly us into the expansive and wild Grosevitchi region where Weyerhaeuser continues to wait for a chance to log. The spruce-forested hills roll unbroken except for fire-born openings on thin taiga soils. I fished bull trout from the pristine waters of the Botchi River and cleansed myself in this icy stream drained from woods where tiger roam with brown bear, Himalayan black bear, wild boar, moose, and sable. I had hoped to spend a week or two in these wildlands, but the vagaries of travel in the Far East, especially during monsoon season, made it likely we would never get out if we didn't cut our visit short.

This is *big* wilderness, set aside not by grassroots organizing or even by fiat, but by bureaucratic inertia. Held against these standards, the wildlands we fight for at home are small (excepting parts of Canada). Compared to the social and political obstacles in Russia and other Third World and repressive nations, America's environmental activists lead sheltered lives of fruitful endeavor. (I am fully aware of the adversaries and tactics we face, from Cointelpro to the wide-abuse movement.) After traveling in Russia, seeing its boundlessness, meeting its activists, no longer do I feel "cutting edge," even as a Wildlands Project big wilderness operative riding the wave of the Pacific Northwest's biodiversity revolution.

This point was further driven home by the smudged-window view from my homeward jet. Clear skies revealed the incomparable majesty of North America's West Coast. The mountains and glaciers of Wrangell–St. Elias punctuated our own wilderness heritage, but continuing down the coast, the diminishing wilderness encroached upon by logging roads and gaping clearcuts added desperation to my tumultuous mental condition. Southeast Alaska, the Queen Charlotte Islands, British Columbia's central coast, all marred by the deep brown of newly exposed soil, evidenced a rate and extent of plunder far in excess of anything in Russia.

We finally flew over my hometown of Bellingham, Washington, unmistakable because of the Georgia-Pacific mill. I strained to see east toward Mount Baker and the mountainous core of the North Cascades ecosystem I have devoted so much to protecting. What used to seem so huge, limited only by the extent of the vision and success of conservationists, now seemed so small, boxed in by cities boiling over and clearcuts spreading like mange.

I think I would rather be a brown bear in the turbulent future of a lawless and undemocratic Russia than in the American landscape of freedom, passion, and efficiency. The future of big wilderness may not be so much a function of the numbers and tactics of its defenders as of the enormous appetite of its insatiable explorers. May some god, any god, protect Russia, not from Weyerhaeuser's chainsaws, but from its corporate culture.

SETTLING IN AGAIN

There is a certain righteous tone, a shrillness, that I think may be distinctly American. I note this undignified tone in the rhetoric of the neo–Sagebrush Rebels, who wisely use media and government to spread their whiney complaints and threats. It is a tone of spoiled immaturity, indicative of a lack of appreciation for some basic truths. By any reasonable standard, Americans are not overtaxed, overregulated, or denied virtually any resource, permit, or law that favors business. The lines against so-called progress are drawn only at the very margins: remnant populations of endangered species, fragments of ravaged forests, wastelands passed over on former binges. Yet even at these margins the shrill whining goes on.

What is it that drives Americans to be so desperately greedy? Maybe it's simply the lingering notion of human primacy and some nervous energy left over from manifest destiny. We are still driving to conquer and pacify the land. Challenging that notion is perhaps the most promising aspect of The Wildlands Project or any plan which displays biocentric thinking. We have far to go. Too many conservationists still cling to aesthetic or recreational justifications for wilderness. They shed one form of resourcism for another.

In my work to protect grizzly bears in the North Cascades, I uncover all too often the dark secrets of what I call consumer environmentalists. This region is more intensively urbanized (and politically progressive) than that of any other grizzly bear ecosystem in the Lower 48. But many of the ranks who write letters for wilderness seem to have in mind only maintaining a safe outdoor playground. One letter written against grizzlies by a stalwart member printed in a conservation newsletter read: "Opposing man's place at the top of the hierarchy is an unlikely stategy for success for the conservation movement; do radical conservationists expect a majority of voters to support operations where man becomes the victim?"

My answer is yes. After watching for ten years the campaign to protect ancient forests unfold, I am convinced that our demands would not have been tolerated had narrow self-interests been foremost. I personally would not favor closing logging towns only to provide views for Sunday drivers or trails for dayhikers. Whatever we have won in the forest fight, it has been through laws that place value on other life-forms and ecosystems and through ethical appeals to respect those rights inherent in life.

America is not a healthy land. Our ecosystems have been plundered and fragmented more than those of less prosperous and less democratic countries, including the other presumed superpower. The great bear roams a mere 1 percent of its former Lower 48 range, and even this is not secure. In Russia, brown bear number in the tens of thousands. The

Wildlands Project makes sense and holds promise because it recalls the proper context and envisions the land restored to health.

What makes The Wildlands Project strategically sound is its therapeutic effect on the human psyche. The vision of The Wildlands Project reinforces an understanding that the natural world has meaning beyond that which we humans give it.

Technology and the Wild

KIRKPATRICK SALE

Technology has always been about the subjugation of nature. Technology is in essence the human construct that is unlike nature, distances us from nature, objectifies nature, allows us finally to manipulate nature. All technology, from the earliest flint to the latest laser, is of a single line seeking control and power over the systems and forces, over the species and treasures, of nature.

What we humans do is simply not in the same league as what other creatures do: our too-large brain has permitted us to construct extrabody implements, nonliving tools, insensate machines do our bidding, the like of which other creatures do not possess or seek to. Perhaps the only way to establish a true, animal, unmediated experience between the human and the wild, therefore, is by the total elimination of technology. It is certainly possible, but not for long, and not as a species. We would not last. Our survival as a species, undeveloped in all muscles except a mind too big, has always depended on tools: tools with which to subdue and survive. *Survive*, at root, means "over life."

The terrible fate that this dependence commits us to is inherently more tragic than any other animal's. It makes technology nothing less for us than a life force, a source of existence. And it makes the development and perfection of technology an inherent part of a desire to improve, enhance, and secure that existence. No wonder we have gone at the business of creating and refining technology with such persistence.

———

Surely at one point there was just the right balance. When? When we were assured of survival by enough artful flints and axes and fishhooks and nooses and baskets, so that we could find enough to hunt and gather for food, enough shelter and warmth for protection. It is, after all, the way we lived on the earth for by far the longest time in our species' history. That can't be by accident.

But at various points, in various remote places, this was not enough technology for those who made the decisions, particularly decisions no longer communal but hierarchical. More technology would mean more power—power over (not merely within) nature—and in some places, for reasons that have yet to be explained, that was thought desirable, at least by a few, and those few dominated. What they created were empires: militaristic, patriarchal, destructive empires that we call civilizations.

And once set upon this path of enlarging human power by the extension of technology, humankind seems not to have been able to halt it, or, perhaps better, seems always to have had among it those few who did not want to halt it. Western capitalism has perfected all that the technological option would allow. No question: if one wanted to imagine a society in which the overt and latent powers of technology could be used to dominate, control, and destroy nature, one could have done no better than to construct the capitalist economy and humanist culture that has developed over the last five hundred years. It has not merely allowed the conquest of nature; it has plotted, sacrificed, and killed to achieve it. Even the continued destruction of vast areas of nature and the extinction of its species were not too high a price to pay in pursuit of this conquest.

And so that magnificent technology has brought the earth—inevitably—to the brink of ecocide and the odds of escape are small. There is not one living earth system that has not been disastrously imperiled in the last fifty years, not one which will be able to survive if technology continues to dictate and humankind to submit. What's worse, each living earth system is the target of some powerful transnational corporation. In the name solely of profit these global corporations pursue courses of action which poison, deplete, exhaust, or eradicate the wild.

There can be no successful human societies without technology. But there can be no successful biosphere without technologies that are carefully limited, proscribed, made simpler, safer, cheaper, smaller, and thoroughly so.

No secret about what that means: the abandonment of all the technologies of modern civilization, the ones *we* use, all of us, and I mean from the telephone to the rifle, the chainsaw to the jet, the computer to the combustion engine.

No, you cannot be selective. It is not merely the machine you buy into when you use the technologies of industrialism, but the worldview of its creators, the mindsets that instruct you not only in how to use it but how to measure its products. These machines do not come ab ovo, unique and separate, out of nothing, but with all the values and limitations of the industrial mind built into them. They are seducing and destroying, twisting your brain cells profoundly, even as they seem to be doing your bidding, saving you time, doing something easier.

Tribal societies know this, which is why they are always resistant to new tools. I am not talking about the decimated and damaged tribes that Europeans encountered in North America—they were merely remnant tribes, frail resemblances of the sturdy pre-Columbian cultures that once prevailed. Even among them the resistance lasted quite some time, in part because Indian technologies, being nature-rooted, were for the most part equal or superior to the first invading technologies. I am talking about healthy, intact, nature-based societies, here in North America and around the world, in which anything new, but particularly any new tool from the outside, is considered very carefully, and in all its multiple meanings, before it is tried out or rejected. As in everything else I know of, our model ought to be the nature-based, tribal societies of the world, the ones whose wisdom, even if we perceive it only imperfectly, is so much greater than any other's. We must ask as they would ask: "What do you have to do, or be, in order to have this tool? What do you choose *against* to choose it? What will it do to your community, not merely of 'two-footeds' but all the rest, not merely now or soon but seven generations hence? What does the community have to say, the elders, the faith-keepers?"

Are we prepared for that—a world where humans abandon the destructive, if daily and seductive, technologies of modern civilization and opt for one where such hard democratic questions are asked and life is lived by imperatives other than the technological? Of course not. Not even the purest of us.

What will it take for us to become so? Most people I know who have thought about this, whose business it is to worry about this, think it will take an ecocatastrophe, a cataclysmic ecospasm of living earth throwing off the civilization that is killing it. They are probably right. But can we be sure we will survive?

Seeing What Isn't There

MARGARET HAYS YOUNG

I live in New York City; I was born here. As a natural area, most folks would agree, New York City has been fairly well destroyed—at least from an ecological standpoint. Growing up in and around New York City, therefore, has given me a pretty good idea of what not to do to an ecosystem.

Other people live in places not so obviously trashed as New York City, or places only recently destroyed, where the damage is fresh and the shame is personal. Such people take a different view of wilderness and react badly to the charge that we have hurt the land. People who live on land that has been ruined within their lifetimes aren't always ready to talk about restoration. In many cases, they aren't even willing to talk about damage. Most especially, they aren't willing to think about responsibility. But it's time now to talk about all these things, from coast to coast.

LOVING THE LAND

I come from a family that has loved nature for a long time. My father took me camping and hiking every summer, starting when I was nine months old. We climbed Mount Washington for the first time when I was four, and climbed it four more times before I was ten. My parents' honeymoon was spent in a tent—three weeks on Cape Hatteras in hurricane season. When I was nine and my brother and sister were six and three, he took us camping for three months around the United States and Canada. We camped with buckets and shovels and tarps in national parks and state forests. We got to know swamps and deserts; we climbed Mount Rainier and the glaciers on Mount Hood, saw the Canadian Rockies and met elk and prairie dogs and buffalo face to face. We saw a lot of this country, and it was wonderful.

My mother's parents were married in Oregon in about 1916, and they immediately set off with packhorses across the Cascades for three months without a map. My grandfather had grown up around a sawmill in Silverton, Oregon. My grandfather's idea of happiness was three months in the wilderness. My grandmother told me her parents were shocked, but she was ecstatic. They were married for fifty years, and at ninety-five she still talked to me about those forests and that honeymoon.

My father moved our family out of the City when I was five, to what was then not yet really a suburb. Over the next ten years, I watched as the woods around our place were razed, and subdivided, and more and more houses were put onto new little lots. We lived in the Hudson Valley, and you could see the river from three sides of our house. When we arrived, there were woodlands all around, with squirrels, inchworms, skunks, rabbits, wildlife everywhere.

But within a few years it was gone, developed into a modern suburb, an empty-spirited, sterile bedroom community. My family hated what was happening, but we couldn't stop it. My father sometimes talked about moving farther north, but his job was in the City, the commute was long, and we knew that the development would follow us soon, no matter how far we went. So we stayed, remembering what the valley had been like, and what my parents had hoped for when they moved there.

On weekends I rode my bike. Old places like Washington Irving's Sunnyside, Lyndhurst, and other prerevolutionary houses gave you a feeling of what the land must have been like throughout the valley two hundred years ago. I could ride thirty or forty miles up the valley on weekends without getting lost, because I could always find the river, and I'd follow the river home. The Hudson is magnificent; its sheer power and strength fill you with awe. It's a mighty river, carving its way through bedrock for millions of years. We went down to its banks as children, but we didn't touch the water. We knew it was polluted and had heard you could get sick and even die if any of it got on you. So we stood and watched it; we skipped snowballs out on the ice in winter, and we loved it, but we knew it was poisoned. I remember wondering what it would have been like if the river were clean. I remember wondering if there were fish in it, and whether they had two heads.

There was a spring near our house, and when I was little, people brought home jugs of water from it; it was clean and fine then. Our neighbors used to talk about how good the spring water was. When we went camping, at first there were streams you could drink from in the mountains; but then we stopped doing that, even in northern New Hampshire and Maine, because "you couldn't be sure" anymore. When we got home from those later trips, I thought about our spring and wondered what you *could* be sure of anymore. Gradually people left the spring alone. We didn't talk about it, people just stopped going.

I think I grew up at a time when the reality of what had happened to the land was just becoming apparent. Camping every summer for a month or more, then coming back to more development in our village every year, I began to have an abiding horror of what was happening. I didn't try to fight it. I didn't think I could—I didn't think *anyone* could.

But as a little kid, I saw nature going under all around me: under blacktop, under landscaping, under huge suburban grocery stores, and, later, under malls. I saw nature going under the gun, as sportsmen came proudly down the roads with dead deer strapped on their station wagons. I saw wildlife going under the wheels of cars on ever-bigger highways. I saw pictures of deformed wildlife, saw pollution-dazzled sunsets and haze-filled sunny days. I read about rain the pH of vinegar and saw trees that looked fried by that rain. And I didn't know what to do. I was a kid. There wasn't anything I *could* do.

I went away to college, and then went away to France. Life was different there, traditions were older; people had built a culture not with plastic and profit, but with art and belief. The cultures we had once come from didn't seem to hate nature quite as much as we did. The land had all been tamed, but it was not the enemy. I came home and finished college, and soon got my first job; I got an apartment, in New York, and I waited. I was twenty. Then suddenly I got lucky; I met and married the man I love. And in time, I began to realize I *could* do what I wanted to do. First, I started doing theater again. I joined conservation groups, Sierra Club, the Museum of Natural History, Audubon, the Wilderness Society, Greenpeace, others, and I started reading. And soon I started writing letters, and then going to meetings, trying to find something effective I could do to help. I think that's how it is with many people who end up working for nature—responsible people sense the destruction and become uncomfortable, unable just to stand by and watch it happen.

WHAT IS NORMAL?

Anyone who feels this way takes on an invisible enemy, because our culture is willfully blind to the effects of development. Like the walls built in Europe against the Infidel, we build walls against Wilderness. We assume that wilderness is what we came here to conquer, what we must defend against. Our assumptions define what we think it is normal for society to do. We have never really questioned our attitude toward wilderness. For five hundred years, we never stopped to reconsider the assumptions that had brought us here and made us believe we had a mission to take, conquer, and then destroy this land.

Our assumptions have been fatal to other species as, ultimately, they will be to us. But they go unnoticed because they're invisible, and you don't question what you can't see. For example, we've always assumed that colonization of this continent by Europeans and others was necessary, beneficial, divinely ordained, and progressive. We assumed that chemical agriculture was scientific advancement and that unending economic growth was the only way to stave off another Depression. We assumed that our spreading suburbs, always building more highways to everywhere, providing more and more cars for everyone, meant prosperity. We accepted "better living through chemistry" and we were sold "our friend the atom." Most people even used to assume that "they wouldn't make it if it wasn't safe." We assumed that more livestock and less wildlife was a good thing for society. We've always assumed that we had a right to be here, to take and "improve" this continent. And we've always assumed that *everyone* ought to have children.

Christian civilizations like the ones that colonized and then spoiled this continent took the biblical directive to "be fruitful and multiply" very much to heart. We've carried on that tradition and have been terribly successful at it. Other species, especially those native here, haven't been as successful at surviving *us*. We've known for a long time that other species are wiped out wholesale by our encroachment on their habitat. But we believed our dominion over the land was ordained by God, calling it our "Manifest Destiny," and to this day many people still quote the Bible to justify our treatment of this land.

When Science became the dominant modern philosophy, we began to call our destruction of wilderness by meaningless names like "scientific wildlife management." A central principle of this notion is the idea that if not "controlled" by us, every species (other than ours) will overrun its habitat, causing mass starvation. Therefore we must "manage" (read: "kill") them—for their own good. Curiously, current scientific wildlife management always seems to lead to more management. An overview of wildlife population trends compared with ours would not suggest that it is *other* species which are breeding out of control.

LEARNING TO ASK THE RIGHT QUESTIONS

It's time to look at what we've done and ask ourselves whether our occupation of this continent has been beneficial. We need to take a look at the land, and see what isn't there. Wherever you are, whatever you're doing, just take a moment to look around: notice what's there, whether it's cars, plastic, glass, steel, asphalt, power lines, TVs or PCs, lawns or trash bins, cows or cockroaches. And then for just a moment, try to see what *isn't* there that might have been there long ago.

Here in New York City, there would once have been huge white pines, giant oaks, elms and chestnuts and tulip poplars. There would have been bears, and wolves, and flocks of birds filling the sky. There would have been free-flowing streams, mosses, marshes, mosquitoes, turtles, butterflies, frogs, owls, and others—and they're not here now. There would have been fish filling rivers you could drink from and clams and oysters at the shore—ones you could eat safely. There would have been native humans ("Indians"), too, not a lot, but some.

Where New York's City Hall now stands, there was once a spring-fed pond. I'm told that the spring is still flowing in the basement, and that the City must keep pumps running to dispose of the water. We need to look at the world now and see what's missing—what we're missing. We don't all have to be experts, we don't need to know every single species of life that's missing, but understanding the land, what it was, what it isn't, will help us to know who we are and to understand what we've done.

Those born into a culture take what they find there as normal because it's all they know. It's hard even to see one's fundamental assumptions, especially when one's friends, family, and neighbors don't even understand the questions. But now it is imperative for us to question our assumptions—as a society and as a species. We might ask whether what we've done is good or sustainable for us. But we ought to ask whether what we have done (and are now selling as a model for progress worldwide) is good or justifiable for anyone to do.

These questions cut across normal political boundaries, and people now joining the fight come from every aspect of the political spectrum. Despite efforts to pigeonhole the environmental movement as leftist or to reduce it to a conventional political argument, nature is not a right-wing or left-wing cause: it supersedes every political construct we recognize. What we have done to the natural world is beyond what any political or social theory can describe. The old words don't apply. Those who speak on nature's behalf carry a new message, previously unknown, politically undefinable. It is a new thing that is happening now.

EAST VS. WEST: WHO KILLED WILDERNESS?

When one questions what has been done by previous generations, the criticism is often most bitterly resented by those who know best that it's true. There's a difference in the way folks in the East and West respond to the charge that we have destroyed this land. In the Northeast and South, where the land has been trashed for many generations, folks are less likely to be defensive about it and more likely simply to admit that it's true. The damage is obvious, but most of it was done a long time ago, by people we never knew.

In the West, the attitude is often somewhat different. There the damage has been done recently, say, within the last hundred and fifty years, and the individuals responsible still have names in living memory. Out West, when one says that the land has been ruined, one may be talking about people's relatives and friends. That's personal. And when it's personal, people get threatened, angry, and defensive. One common reaction is to deny that there is a problem, however obvious it may be. That's happening now: many folks will stubbornly stand on overgrazed, eroded, compacted, clearcut, or poisoned land and simply insist there's nothing wrong with it.

In the East, most of the destruction was accomplished many generations ago, and while the damage is worse, it's more anonymous. In the East, especially in the cities, few people question the obvious fact that ecosystems have been destroyed, but we don't have to know who did it. Sure, in some cases they were our ancestors, but we don't know which ones; it all happened such a long time ago. We live with the consequences, the fouled air, the poisoned rivers, the absent wildlife, the endless toxic dumpsites, the absence of an "outdoors" or a sky in the barren long canyons of high-rises. But *we* didn't do it, it was done *to* us, by people we never met.

If you suggest to a public lands rancher, for example, that cows and sheep are destroying the land, he often bridles and denies it. Perhaps because he knows how it happened and doesn't want to be held responsible. Perhaps his parents or grandparents have been given public tax dollars to do it for a long time. Perhaps, as a western rancher, he takes a certain pride in being part of the heroic Western Lifestyle. When a public lands logger in the Northwest insists that "forests grow back," though he may see the clearcuts expanding all around him and knows that big trees are getting hard to find, he's caught in a quandary: he can see that the land is ruined, but he's still got the chainsaw in his hand.

When a logger or rancher or miner insists on his right to his traditional lifestyle, he's really insisting on three things: that western settlers had to occupy the land for the Good of the Country; that they therefore had a right to public subsidies to keep doing it; and that they are part of an American Tradition which the rest of the country admires. If he is mistaken, then his tradition may have no meaning and his livelihood may be in serious jeopardy. Western settlers weren't just ordinary colonists: they thought they'd been given a Mission, they thought they had a Deal, and they thought their traditions were heroic. But suddenly people are saying that those are just empty myths.

While easterners and Californians have renegotiated their expectations, many western folks have not. Because the land has been ruined, a lot of myths about the West are being questioned all at once. There is more to western resentment of eastern "meddling" in wilderness issues than resistance to new ideas. To a public lands user, these new ideas may cost him not only access to the land, and some long-accustomed subsidies, but a lot of old myths and a great deal of self-respect. In such circumstances, it is not surprising that many westerners feel cheated and betrayed. The early western settlers carried a nineteenth-century mandate which, in retrospect, looks a lot like a mistake. It was perhaps best said by a Colorado feedlot owner quoted by Anne Matthews in her wonderful book *Where the Buffalo Roam*: "Asking us to admit that we were wrong all along, in trying to settle a lot of this country, is like asking us to have surgery without anesthetic." This statement has resonated for me ever since I first read it. It is about asking folks to admit they were wrong, and it is that admission which is so painful, and so intolerable.

WHO'S TAKING WHAT?

We didn't come here with the intention to destroy the land. We did, however, come here looking to get rich, and it has been the process of getting rich that has caused the damage. As the damage becomes more obvious, the guilt associated with getting rich by destroying what we love becomes intolerable. That guilt, and our own hypocrisy, has caused many people to retreat into denial. Faced with the prospect that Western Expansion and colonization of this continent may not have been noble or heroic at all, we are looking for some way out, some way to believe the old myths again. But the land is witness to what we have done here, and the land doesn't lie.

The anger and sense of betrayal many westerners feel is being deliberately fanned by a growing collection of groups calling itself the "Wise-Use Movement." Allied with other right-wing groups, including the New Christian Right, and quietly funded by large corporations, they are mounting an aggressive, extremely well organized propaganda campaign to frighten and organize rural people against the environmental movement. Their published "Wise-Use Agenda" advocates opening up all public lands, including national parks, to unlimited resource extraction and overturning all environmental laws. The twin rallying cries of this campaign are "Property Rights!" and "People First!" They call upon rural folks to defend their "traditional lifestyles," which they claim are being taken away by environmentalists. The fact that these threats can be persuasive to so many people is an indication of how frightened and irrational we have become. For centuries, people on this continent have always come first; the rights of people have been all we've ever considered. And as a result, much of the native wildlife is gone. There has never been any lack of putting people first; we have never done anything else.

Creating a perceived threat to individual property rights has been an efficient way to frighten people. Our current "property" was granted to our forebears by various kings in Europe. But what exactly are our rights to the property we occupy and claim to own? One might ask how any European king could make grants of land on a continent he had never seen. Perhaps we are nervous about our property rights because we aren't quite able to explain how we got them.

We are here by right of occupation and by no other right. Our forebears negotiated treaties with Native tribes, still legally in force, which we have violated systematically. Wise-use groups advocate using the "takings" provision of the Constitution to prevent environmentalists from protecting land and wildlife. Amendment V states: "No person shall . . . be deprived of life, liberty or property, without due process of law, nor shall private property be taken for public use, without just compensation." Does this provision apply only to white people? Does it apply only to people? If we have subsidized settlers to ruin the land, who has taken what from whom? And yet these arguments have been very seductive to a lot of threatened, frightened people who would do anything to avoid admitting that what we have done here was wrong.

Hypocrisy has defined our national character, since this nation was formed, and it has bedeviled us ever since. We claim that we came and settled in an open and unclaimed country, that we improved it by developing it. What we claim is not true. We stole this land, and we have largely destroyed it. We can choose to admit this, however painful it may be, and try to put back some of what we have taken. Or we can choose to keep lying about it and continue to destroy the land.

What if we keep on denying it? How long can we hold out? The evidence is all around us; we can see it and smell it and taste it; it isn't getting better and it isn't going away. If we keep on telling ourselves that "forests grow back," we'll still see barren moutainsides and mud-filled rivers. Soon there won't be any salmon left at all. Shall we just keep on denying it? How long can we keep it up? We're about to lose the wolf and the grizzlies forever. Shall we just pretend we don't remember them, once they're gone?

The trouble with denial is, it isn't going to work. The problem with the wise-use agenda is that while it may be comforting for a little while to pretend there isn't a problem, the proof of what we've done is written on the land. We can face it now, or we can face it later. For decades, we've been opting to face it later. The longer we put it off, the worse it will be when we finally have to tell the truth.

Protected Areas
in Canada

JAMES MORRISON

*Whereas Canada's aboriginal peoples hold deep and direct ties to
wilderness areas throughout Canada and seek to maintain options for
traditional wilderness use . . .*
Preamble to the *Canadian Wilderness Charter*

The Canadian Wilderness Charter's eloquent wording suggests that
conservationists and aboriginal people share common aims and objectives with regard to protected areas. The general public certainly sub-
scribes to this view, one easily reinforced by recent events—from the
struggle for South Moresby and the Oldman River to the battle over
old-growth timber in Temagami.

But as Georges Erasmus points out in his contribution to the book
Endangered Spaces, aboriginal interests are not identical to those of the
conservation community. For the Native leadership, at least, wilderness protection is only one part of a larger political question—one
"bound up with the thorny issues of treaty rights, aboriginal title and
land claims." The indigenous people of Canada, says the former Grand
Chief of the Assembly of First Nations, are seeking not only recognition
of their inherent right to govern themselves but also a land and resource
base adequate to support their communities (Hummel 1989).

Nowhere has the issue been joined with more fervor than in Ontario.
Many people in the conservation community reacted first with surprise
and then outrage when—as part of land-claim negotiations—the new
provincial administration announced that game legislation would not

This is an edited and abridged version of a discussion paper originally published by
World Wildlife Fund Canada in July 1993.

be enforced against members of the Golden Lake Nation found hunting within the bounds of Algonquin Park. A short time later, Ontario entered into negotiations with the Lac la Croix Ojibway of northwestern Ontario, who were seeking increased motorized access to Quetico Park for fishing purposes. This too sparked anger.

Many of the perceived differences between a conservationist view of protected areas and one based on aboriginal rights were clearly summarized in an exchange of correspondence about Quetico in the Toronto *Globe and Mail*. On 18 May 1992, journalist Robert Reguly accused the Ontario government of giving the Lac la Croix Ojibway privileges which violated the park's status as a protected area. Like many wilderness advocates, he particularly objected to opening up the park to motorized travel. Law professor Kent McNeil was quick to respond. He argued that the creation of Quetico Park had actually violated an 1873 treaty with the local Ojibway by excluding them from hunting and fishing within park boundaries. Canadians, he said, ought to reflect on the fact that only 0.3 percent of the country had been set aside for indigenous people. "I am not against the creation of parks or wilderness areas, but surely the few rights the aboriginal people have left should take precedence over the pleasure of canoeists and campers." This was too much for Kenneth G. Beattie of Toronto. Accusing Professor McNeil of shallow thinking on aboriginal rights, he insisted that the Lac la Croix people simply wanted expanded access to the park because they had depleted fish stocks elsewhere—much as Ojibway people had already destroyed the Winnipeg River sturgeon fishery. Treaties, he argued, should be interpreted in the light of modern principles of resource management—for "uncontrolled exploitation of natural resources results in the destruction of those resources, regardless of the racial origin of the exploiters."

The actual or potential conflict between these positions will have major consequences for the Endangered Spaces Campaign. World Wildlife Fund Canada's visionary goal of increasing the number of protected areas in all of Canada's natural regions will inevitably be caught up in the constitutional crossfire over Native self-government. Not only are some proposed spaces likely to fall under Native jurisdiction, but more and more parks and protected areas throughout the country will become the subject of claims to aboriginal or treaty rights. Sorting out the questions of jurisdiction and title will slow governmental action on new protected areas—and make it that much more difficult to complete the Endangered Spaces agenda by the year 2000.

If it is no longer possible to ignore the differences between conservationists and aboriginal people, is it still possible to ensure the protection of vanishing wildlife and wilderness areas? This essay searches for common ground. First we need to examine what Georges Erasmus calls the profound philosophical cleavage in cultural points of view between indigenous and nonindigenous people in Canada. These differences have a history. If they are not understood and addressed, then long-dormant hostilities could overwhelm efforts on both sides to protect endangered spaces.

A WILDERNESS ETHIC

In the Aboriginal worldview of the Four Orders, the Aboriginal person is viewed as last: this is in acknowledgment of the natural superiority of Manitou, Earthmother, the Plants and Animalkind. From this subservient position the Aboriginal person is imbued with a sense of the sacredness of all things as gifts and manifestations of a benevolent and caring Manitou.

—CECIL KING, ODAWA TEACHER

Wilderness, in contrast with those areas where man and his own works dominate the landscape, is . . . an area where the earth and community of life are untrammeled by man, where man himself is a visitor who does not remain.

—U.S. WILDERNESS ACT, 1964

Although there is no single definition of wilderness, many conservationists would acknowledge the philosophy expressed in the first great piece of American wilderness legislation. These ideals were popularized at the turn of this century by conservationists like John Muir, who argued that there had to be spaces free from urbanization and industrial development, where the human species could recognize its own insignificance and retain a sense of awe at the wonders of creation.

Muir was reacting to—and rejecting—the modern conception of

nature as an enormous reservoir of energy and resources that the human race can dominate and exploit with impunity. In this he had much in common with the views of indigenous societies who have consistently placed humankind in a subservient position to the rest of creation. To aboriginal people, says Cecil King, an Odawa educator from Manitoulin Island in Ontario, the idea that humans are a superior species who can dominate the natural world is blasphemous. Indeed, as more and more people worldwide now realize, it is the modern conception of nature which has led to the destruction of our environment and threatened the very survival of our own and other species.

But despite their apparent similarity, there are fundamental differences between indigenous and nonindigenous conceptions of nature. As poet and naturalist Gary Snyder has explained, one thread of the conservation movement is profoundly romantic in that it sees the human species as an intruder, not as part of the natural world. Protected area management on this continent has tended to reflect that philosophy. Canada's national parks policy, for example, speaks of protecting and managing natural resources in parks "to ensure the perpetration of naturally evolving land and water environments and their associated species." The expression "associated species" does not necessarily include humans.

By contrast, indigenous societies both past and present place mankind at the axis of the natural world—subordinate to the whole, but essential. Nowhere was this more apparent than among the pre-Columbian Olmecs and Aztecs of Mesoamerica. There, Mexican poet Octavio Paz tells, humanity's role was the giver of blood. It was human sacrifice which drove the world, enabling the sun to rise and the corn to grow. If less terrifying in import, similar cosmologies have prevailed in all aboriginal cultures. Without proper offerings to show respect for the spirits—or what Cecil King calls the *manitous*—hunts will fail, the fish will vanish, and the universe will come to a halt.

To indigenous people, wilderness—in the sense of areas "untrammeled by man"—does not exist. Geographer Peter Usher, a pioneer in the field of Native land-use and occupancy studies, has shown that the wildest parts of Canada are far from being empty spaces. Even if they appear to be underutilized, they are occupied by indigenous people on the basis of detailed knowledge going back hundreds or even thousands of years. Graves and habitation sites dot the landscape. The mountains and hills, lakes and streams, trails and portages, all have names and legends associated with them. This is as true today for Micmac and Malecite fishers on the Restigouche River in Quebec and New Brunswick—who have been in continuous contact with Europeans since the sixteenth century—as it is for the Inuit or Déné hunters in the remote arctic and subarctic regions of the country.

At the core of the indigenous relationship with nature is a reciprocal connection with the plant and animal world. Because of this, many aboriginal people share with conservationists what can reasonably be called a wilderness ethic. A clear, deep, spring-fed lake is as positive a value to an Ojicree trapper in northeast Manitoba as it is to a recreational canoeist from Winnipeg. And an eagle is as worthy of respect and awe—both for its innate beauty and for its connection with the thunderbird of Native legend. In aboriginal communities across Canada, physical well-being is closely associated with nature. "Country food" such as wild fish and game is uniformly perceived as healthy, store-bought food as unhealthy.

Like many indigenous leaders today, Georges Erasmus insists that Native peoples "have a keen interest in preserving areas as close as possible to their original state" (Hummel 1989). After all, they have experienced the alternative. Mississaugas living on the New Credit Reserve near Brantford remember Etobicoke (Adoopekog) as the "place of the alders" near Lake Ontario. What was part of Mississauga territory in the mid-nineteenth century is now a suburb of Toronto. The alders are gone, the lake and creek are polluted, and the fish no longer thrive.

Without renewable resources to harvest, as Georges Erasmus puts it, aboriginal people lose both their livelihood and their way of life. That way of life is not a folkloric remnant. In his latest book, former B.C. Supreme Court Justice Thomas Berger argues that most Canadians misunderstand the Native "subsistence" economy. Because our world is industrial, we tend to see aboriginal people as anachronistic. Either Natives are living a precarious existence on the edge of starvation and must be weaned into the mainstream economy or, a view held by many environmentalists, they should be permitted to continue their subsistence activities, provided they adhere to "traditional" methods and patterns of harvest.

The second view is certainly more benign. The first—which sees the Native economy as "unspecialized, inefficient, and unproductive"— has, as Berger claims, resulted in enormous social upheaval. During the 1950s, especially in the north, governments evacuated aboriginal people from their habitual territories and resettled them in new villages in the hopes that wage and salaried employment would eventually be provided. Those jobs, with few exceptions, have never materialized, and likely never will. The alternative is an economy based on hunting, fishing, and trapping supplemented by occasional wage labor or transfer payments. In the north, such an economy remains traditional in the sense that it continues to bind people together in an older web of rights and obligations. In the Cree communities of eastern James Bay, the best hunters still enjoy the greatest social prestige and game or fish are distributed according to age-old patterns (Scott 1986).

In one important respect, however, the subsistence economy is anything but traditional. Thomas Berger points out that Native people everywhere now use outboard motors, snowmobiles, or all-terrain vehicles (ATVs) in their hunting and fishing activities—much as in earlier generations they adopted canvas canoes and muskets in the place of bark or skin boats, spears, and bows. Déné from northern Saskatchewan even fly into the Northwest Territories to hunt caribou rather than travel overland by canoe or snowshoe. Indigenous people do not share the antipathy felt by many in the conservation community toward technology—including mechanized forms of wilderness travel—since boats or snowmobiles are not really used for recreation. These modern devices simply make it easier to earn a living.

For their part, conservationists raise legitimate fears about the long-term effects of new technology on wildlife survival. This is the real nub of much of the current conflict between the two sides. Do modern methods make it easier to harvest—and therefore threaten or eliminate— wildlife species? Such concerns appear to be reinforced by demographic trends. By all estimates, Native people have the highest birth-rate in Canada. On Indian reserves across the country—in marked con-trast to the aging general society—children and adolescents now constitute the largest single population group. Assuming that tradi-tional harvesting continues at the same rate, then a larger Native pop-ulation could put added pressure on fish and wildlife species. This observation should be balanced against another social trend. Over the past few years, there has been an astonishing rate of aboriginal migration from rural to urban centers—not only to large cities like To-ronto, Winnipeg, Regina, and Vancouver but to smaller centers in most regions as well. In Ontario alone, some 40 percent of Native people already live off-reserve and this number is growing rapidly (Bobiwash 1992). This trend too is of concern to some conservationists. In virtually all urban areas—as well as in many rural or northern Native communities—young aboriginal people have lost or are losing tradi-tional bush skills. Without the wilderness ethic of their elders, would aboriginal people continue to show respect for wildlife?

Despite such questions, aboriginal people are only a small part of the perceived problem. Most of the anger and frustration voiced by con-servationists is related to the diminishing supply of wild places throughout Canada. Urbanization is an obvious target—as the struggle to preserve the Rouge River valley in suburban Toronto has shown. But the lack of planning and development controls in rural municipalities has also led to the destruction of unique vegetation and wildlife habitat, as has the inexorable march of industrial development on Crown lands. In much of southern Ontario, to give the most prominent example, there is no longer sufficient wild country to allow for the creation of fully representative protected areas (Hackman 1992).

Against this background, aboriginal issues can be seen either as a distraction or as a luxury. In a recent volume celebrating the centenary of the Ontario parks system, John Livingston vigorously attacks the ideology of human proprietorship over nature. In the contemporary discussion of Native claims, both aboriginal people and different levels of government consistently focus on the "management" of wildlife "re-sources" as a primary goal. Management, he notes bitterly, "is the usual euphemism for deciding on what numbers of what species of living beings may be killed, where, when, by whom, and by what means" (Livingston 1992).

Rather than concentrate on perhaps irreconcilable policy differences, conservation groups like World Wildlife Fund Canada have devoted much of their energy to counting and monitoring wildlife populations. As part of this goal, however, they too seek answers from aboriginal people and their political organizations. Echoing John Livingston, they ask whether an apparent fixation on treaty and aboriginal harvesting rights leaves any room for conservation. This concern has been sparked by disturbing recent events. In 1992, to give a prominent example, Fisheries and Oceans Canada agreed to recognize an exclusive Native food fishery along the Fraser River in British Columbia. While some First Nations complied with their own laws or governmental regulations—and indeed counted and monitored fish populations—other Native people along the Fraser have been accused of transporting large quantities of fish to markets in the United States.

Aboriginal people have not responded directly to these issues. There are several reasons why they have tended to concentrate on questions of title and rights. For one, their experience with the creation of parks and protected areas, as well as with the enforcement of fish and wildlife regulations, has made many of them deeply skeptical of the goals and motives of both government and the conservation movement. Too often over the past century, say Native leaders, governments have either ignored or violated their aboriginal and treaty rights—sometimes at the urging of conservationists, who have cited the same kinds of concerns about aboriginal harvesting practices.

TREATY AND ABORIGINAL HARVESTING RIGHTS

And the said William Benjamin Robinson of the first part, on behalf of Her Majesty and the Government of this Province, hereby promises and agrees . . . to allow the said Chiefs and their tribes the full and free privilege to hunt over the territory now ceded by them, and to fish in the waters thereof as they have heretofore been in the habit of doing, saving and excepting only such portions of the said territory as may from time to time be sold or leased to individuals, or companies of individuals, and occupied by them with the consent of the Provincial Government.

—ROBINSON TREATIES, 1850 (NORTHERN LAKES HURON AND SUPERIOR)

The existing aboriginal and treaty rights of the aboriginal people of Canada are hereby recognized and affirmed.

—CONSTITUTION ACT, 1982, SECTION 35(1)

Aboriginal people are the only sector of Canadian society who possess constitutionally recognized—and protected—rights to harvest fish and wildlife. This reality does not sit well with the animal rights movement. Nor does it appeal to modern sportsmens' groups such as the Ontario Federation of Anglers and Hunters. Although the latter usually cite the impact of Native rights on conservation policies, in reality their disagreement is more fundamental. To them, Native harvesting rights are undemocratic because they confer special privileges on one group of people. This opinion is widely shared by non-Native people in rural and northern areas of the country.

The idea of one Canada for all Canadians—whatever their origins— was popularized by former Prime Minister Pierre Trudeau, who fully intended that the concept be expanded to aboriginal people. In a White Paper published in 1969, the Trudeau government proposed abolishing the walls separating Native people from the rest of society—largely symbolized by the Indian Act—and transferring program responsibilities to the provinces. To Trudeau, it was unthinkable that one sector of society should have treaties with another. Bringing Native people into the mainstream would help solve the problems of poverty and powerlessness that were such glaring social problems.

The virulence of the Native reaction to these proposals took the government by surprise. Rejecting assimilation as a product of Western theories of racial superiority, aboriginal people argued that the Indian Act—though a colonialist document—was still a testimony to the direct and special relationship they had always enjoyed with the Crown. This relationship entirely bypassed "white settler" governments— which were represented, after confederation, by the provinces. Aboriginal people made it clear that they sought their own governing institutions within confederation, institutions which would be parallel, not subordinate, to provincial governments (Marule 1978).

To aboriginal people, treaties embody the special relationship between themselves and the Crown. In their view, these agreements symbolize the fact that Canada is not simply a settler society—but is instead linked formally to the distinct aboriginal societies that were here when the Europeans first arrived. Harvesting rights are an important part of that special relationship and are still integral to most aboriginal societies in Canada. This helps explain the tenacity with which Native groups have, over the past two decades, fought to have those rights respected.

PARKS AND PROTECTED AREAS

The creation of most early protected areas in North America involved the exclusion of aboriginal people. The most obvious example is Yellowstone Park. The park was established by Congress in 1872 in the midst of the post–Civil War campaign to subdue the Sioux and other Plains Indian tribes. The inhabitants of Yellowstone—mainly Crows and Shoshones—either left for reservations or were driven out by the U.S. Army, which would manage the park until 1916.

In Canada, government regulation took the place of force of arms. Protected area management took more than one form in the early decades of the twentieth century. Both Quetico and Riding Mountain parks were originally set apart as forest reserves. Many provinces also created game preserves within their jurisdictions. Aboriginal people, however, did not distinguish between the various categories, since most had similar impacts on their harvesting rights. In 1925, for instance, Ontario banned all hunting and trapping within the Chapleau Crown Game Preserve, a tract of several thousand hectares in northern Ontario. Not only did the action permanently affect the livelihood of a few hundred Ojibway and Cree people, it forced one of the bands to surrender its reserve in the center of the game preserve. Ironically, those lands were eventually incorporated into Missinaibi Lake Provincial Park in the early 1970s. In 1974, the National Parks Act was amended to recognize aboriginal hunting, fishing, and trapping in parks or park reserves north of the 60th parallel. But with the exception of Pukaskwa in Ontario, the same recognition has not been extended to southern properties. Until recently, provincial jurisdictions have generally refused to consider such Native access to parks and protected areas.

North American parks and protected areas have generally been created in the name of the public interest. Most conservationists fully support this concept—insisting that it is a governmental responsibility to protect significant regions of the country for the benefit of future generations. Aboriginal people, however, dispute the inclusiveness of the term "public." In their view, it automatically places the interests of the general society above those of minorities. They point out that governments also cited the public interest when imposing large-scale resource development projects on them—such as pipelines and hydroelectric dams. As the examples of the Oldman Dam and James Bay II hydro projects show, governments continue to use the same arguments today.

It is fair to say that indigenous people have borne the costs of protecting natural areas, through the loss of access for hunting, trapping, or other harvesting activities. As Georges Erasmus puts it, the doctrine of the public interest made "an ancient way of life subject to the apparent modern-day whims of alien culture, all in the name of conservation" (Hummel 1989). Conservationists, nevertheless, vigorously defend the parks system. While they may concede a certain lack of historical sensitivity or understanding, they argue that the fundamental choice was never between protected areas and aboriginal interests, but rather between protection and industrial development. If anything, they say, the situation would have been infinitely worse without the skilled polemical lobbying of groups like the Algonquin Wildlands League and the Canadian Parks and Wilderness Society. The scale of clearcut logging or mining on Crown lands would have been far more significant and the damage to the habitat of fish and wildlife sought by aboriginal people that much more severe.

In this kind of dispute, Canada is far from unique. The same arguments, and the same tensions, over parks and protected areas are being worked out in various parts of the world. And in at least some jurisdictions, the lessons being learned are positive ones.

TOWARD A SOLUTION

Native people do not want to recreate a world that has vanished. . . . They do not wish to return to life in tipis and igloos. They are citizens of the twentieth century. However, just because Native people use the technology of the dominant society, that fact does not mean that they should learn no history except that of the dominant society, or that they should be governed by European institutions alone. Native people want to develop institutions of their own fashioning; they are eager to see their cultures grow and change in directions they have chosen for themselves.

—THOMAS BERGER, 1991

It has always been part of basic human experience to live in a culture of wilderness. There has been no wilderness without some kind of human presence for several hundred thousand years. Nature is not a place to visit, it is home—and within that home territory there are more familiar and less familiar places.

—GARY SNYDER, 1990

Regardless of its exact contours, a new level of authority—Native self-government—is being created in Canada. When it is complete, the aboriginal people of Canada will arguably have more powers than any other indigenous group in the world. While some members of Canadian society envision the creation of ethnic enclaves and fear the potential destabilization of the body politic, the recognition of the inherent right to self-government will have a particular impact on relations with the provinces and territories. While most aboriginal people see themselves as Canadians—and as Nisga'a, Inuit, or Cree—few have ever regarded themselves as citizens of British Columbia or Ontario or Nova Scotia. To them, provincial governments have always been the representative institutions of non-Native settlers. This outlook explains the frequent insistence by aboriginal leaders that talks now be conducted with them on a "government to government" basis. They see their eventual self-governing institutions as being at least parallel in status to those of the provinces. Ontario was the first jurisdiction to formally acknowledge this fact, in the August 1991 Statement of Political Relationship.

The federal government and many provinces argue that Native self-government will apply only to existing Indian reserves or community lands created through negotiation. This is definitely not the view of aboriginal political leaders. They point out that the 1982 Constitution Act already recognizes and affirms their aboriginal and treaty rights. As several recent Supreme Court decisions have made clear, these rights include priority of access to unoccupied Crown lands and waters for hunting, fishing, trapping, and other subsistence pursuits. Governments may well have difficulty maintaining the view that parks and protected areas are "occupied" Crown lands which prevent the exercise of treaty and aboriginal rights.

Claims settlements in the Northwest Territories and Yukon already acknowledge aboriginal interests in archaeological sites and areas of cultural or spiritual significance on nonsettlement lands. This is a trend in provincial heritage legislation as well, one which will give Native people a say in the management of lands off their reserves or settlements. Parks and protected areas are bound to be included in this category.

The combination, therefore, of self-government initiatives, the settlement of Native claims, and constitutional recognition of treaty and aboriginal rights will have an obvious impact both on the Endangered Spaces Campaign and on protected area management in general. As we have seen, this has already happened in some areas—particularly those under federal jurisdiction, such as the Yukon and Northwest Territories. The results there have been largely positive. Both Canada and aboriginal groups have agreed to provide for new protected areas as part of land claim settlements. The Inuvialuit, Inuit, and other groups in the Yukon and Northwest Territories have also ensured that their interests—including employment opportunities and cultural survival through continued harvesting rights—are fully protected as well.

Within the provinces, however, the situation is much more problematic. It has been hard enough to reach land claim settlements between aboriginal people and the federal government alone. The addition of the provinces and private or third-party interests make agreements that much more difficult. And in provinces like Ontario, the unrelenting hostility of angler and hunter groups to treaty and aboriginal rights has greatly complicated the task. Virtually all of British Columbia is or will be subjected to Native claims of one sort or another.

The same is true of large portions of Quebec and the Maritimes, which also have few, if any, treaties with their Native inhabitants. The prairie provinces are blanketed with claims based on treaty entitlement. A September 1992 announcement in Saskatchewan commits the three levels of government—Saskatchewan First Nations being the third—to negotiate and settle the considerable areas of land still owing to aboriginal people under treaties signed between 1874 and 1910. Some lands with conservation potential may well be selected.

In much of the country, therefore, it is the Native political agenda which will influence protected area programs, not the other way around. In northern Quebec, for example, the Cree are interested in park proposals because they see them as one means of halting James Bay II. Their kinsmen in western James Bay have similar goals for the Moose River basin. If parks or protected areas will prevent further hydroelectric development on the rivers leading to the bay, then they are in favor. The same is true of Montagnais from the lower north shore of the St. Lawrence. They too want to stop hydroelectric development in their traditional homeland. In other regions of the country, it is possible that the claims process will slow, rather than speed up, progress toward the year 2000 target date for the Endangered Spaces Campaign. This already seems to be happening in British Columbia, where tensions between aboriginal groups and environmentalists—exacerbated by government decisions about areas like Clayoquot Sound and Tatshenshini—are rising as each side pursues its own goals.

These tensions flow from the differing viewpoints summarized in the course of our discussion. And they are highlighted in the exchange of letters in the *Globe and Mail* with which we began. It is therefore worth posing the questions again. Does allowing aboriginal people motorized access violate the fundamental protected status of parks, as Robert Reguly argues? Should the "few rights the aboriginal people have left," to quote Kent McNeil, take precedence over the pleasure of canoeists and campers? Or should treaties be interpreted, according to Kenneth G. Beattie, only in the light of modern principles of resource management?

So long as the argument pits Native subsistence against the recreational needs of an urbanized general society, aboriginal people will always have the moral upper hand. Indeed, their message to conservationists and to government is that without respect for existing treaty and aboriginal rights, new conservation agreements will not be possible. They will no longer allow their rights to be sacrificed on the altar of some larger public interest. In the planning of new protected areas, aboriginal people also expect to be involved from the very beginning. They want protected area managers to realize that their participation in planning and management is not a threat but a guarantee of their own livelihood and a positive contribution to the preservation of wild spaces.

But Kenneth Beattie raises another issue when he warns that uncontrolled exploitation of natural resources results in the destruction of those resources "regardless of the racial origin of the exploiters." Wild spaces everywhere are shrinking or vanishing altogether before relentless human pressure. Even in a country as thinly populated as Canada, some plant and animal species have already become extinct and many more are vulnerable. Conservationists do not accept that treaty and aboriginal rights should be an end in themselves—despite what Professor McNeil has to say—particularly if they substitute one type of human predator for another. In southern Canada, faced with the twin pressures of urbanization and industrial development, conservationists continue to see aboriginal interests as secondary to the ultimate survival of wildlife and natural areas.

Conservationists and aboriginal people should not be expected to agree on goals or tactics. What is needed is respect for divergent positions so that gains can be realized. While the two sides may differ on their ultimate objectives, they do have common interests. One is a shared antipathy to the type of large-scale resource development which has ravaged much of Canada. Another is a commitment to the ethos expressed recently by naturalist Ron Reid—that while humans have become the dominant species on earth, "we still have the cardinal responsibility to share our whole planet with all other living creatures, plant and animal, that evolved here."

The advantages of common action far outweigh the disadvantages.

Even reluctant provincial governments can be persuaded to come along. To paraphrase a saying commonly used by aboriginal leaders: both groups do not have to travel in the same canoe. But they can certainly share a waterway and even arrive at a common destination.

References

Berger, Thomas. *A Long and Terrible Shadow: White Values, Native Rights in the Americas, 1492–1992*. Toronto: Douglas & McIntyre, 1991.

Berringer, Patricia. "Aboriginal Fishery Systems in British Columbia: The Impact of Government Regulations, 1884–1912." Paper presented to the 16th Annual Congress of the Canadian Ethnology Society, Ottawa, 1987.

Bobiwash, Rodney. "The Provision of Aboriginal Social Services in a Native Urban Self Government Paradigm." In B. Hodgins et al., eds., *Co-Existence: Studies in Ontario–First Nations Relations*. Peterborough, Ont.: Trent University, 1992.

Hackman, Arlin. "The Job to Be Done." In L. Labatt and B. Littlejohn, eds. *Islands of Hope: Ontario's Parks and Wilderness*. Toronto: Firefly Books, 1992.

Hummel, Monte, ed. *Endangered Spaces: The Future for Canada's Wilderness*. Toronto: Key Porter Books, 1989.

King, Cecil. Review of Angus, "And the Last Shall Be First." *Compass: A Jesuit Journal* 10(4) (Sept.–Oct. 1992):42–43.

Livingston, John. "Attitudes to Nature." In L. Labatt and B. Littlejohn, eds., *Islands of Hope: Ontario's Parks and Wilderness*. Toronto: Firefly Books, 1992.

Lytwyn, Victor P. "Ojibwa and Ottawa Fisheries Around Manitoulin Island: Historical and Geographical Perspectives on Aboriginal and Treaty Fishing Rights." *Native Studies Review* 6(1) (1990):1–30.

Marule, Marie Smallface. "The Canadian Government's Termination Policy: From 1969 to the Present Day." In A. L. Getty and D. Smith, eds., *One Century Later: Western Canadian Reserve Indians Since Treaty 7*. Vancouver: Pica Press, 1978.

Miner, Jack. *Jack Miner: His Life and Religion*. Kingsville, Ont.: Jack Miner Migratory Bird Foundation, 1969.

Morris, Alexander. *The Treaties of Canada with the Indians of Manitoba, the Northwest Territories and Keewatin*. Toronto: Willing & Williamson, 1880.

Paz, Octavio. "The Power of Ancient Mexican Art." *New York Review of Books* 37(19) (1990):18–21.

Price, Richard, ed. *The Spirit of the Alberta Indian Treaties.* Vancouver: Pica Press, 1987.

Reid, Ron. "Ontario Parks: Islands of Hope," In L. Labatt and B. Littlejohn, eds. *Islands of Hope: Ontario's Parks and Wilderness.* Toronto: Firefly Books, 1992.

Scott, Colin. "The Socio-Economic Significance of Waterfowl Among Canada's Aboriginal Cree: Native Use and Land Management." Paper presented to the 19th International Council for Bird Preservation, Kingston, Ontario, 1986.

Snyder, Gary. *The Practice of the Wild.* San Francisco: North Point Press, 1990.

Tough, Frank. "The Establishment of a Commercial Fishing Industry and the Demise of Native Fisheries in Northern Manitoba." *Canadian Journal of Native Studies* 4(2)(1984):303-319.

————. "Ontario's Appropriation of Indian Hunting: Provincial Conservation Policies vs. Aboriginal and Treaty Rights, ca. 1892-1930." Unpublished report prepared for Ontario Native Affairs Secretariat, 1991.

Usher, P. J. "Indigenous Management Systems and the Conservation of Wildlife in the Canadian North." *Alternatives* 14(1)(1987):3-9.

Usher, P. J., F. Tough, and R. Galois. "Reclaiming the Land: Aboriginal Title, Treaty Rights and Land Claims in Canada." *Applied Geography* 12 (1992):109-132.

Van West, John J. "Ojibwa Fisheries, Commercial Fisheries Development and Fisheries Administration, 1873-1915: An Examination of Conflicting Interest and the Collapse of the Sturgeon Fisheries of the Lake of the Woods." *Native Studies Review* 6(1)(1990):31-65.

World Wildlife Fund. *Whales Beneath the Ice: Final Report, Conclusions and Recommendations Regarding the Future of Canada's Arctic Whales.* Toronto: World Wildlife Fund, 1986.

Restore Wildness

BILL DEVALL

Fog blows across Rodgers Peak, settling into the unnamed tributary of Redwood Creek in the late afternoon. We are walking along a "restored" slope of this watershed in Redwood National Park. In the early 1970s a timber company bulldozed a major road across these slopes in order to remove ancient redwoods by clearcut logging. In place of ancient redwoods, fifteen years later, are thick stands of alder and young Douglas fir.

When Congress expanded Redwood National Park in 1978 to include mostly clearcut lands, it mandated a restoration program. A survey by the National Park Service in 1979 found over three hundred miles of logging roads and at least a thousand miles of skid trails inside the expanded park boundaries.

A supervisory geologist explains the process of road building and road removal to a group of students who are interested in becoming restoration workers. "We're looking at the largest project we've ever taken in road removal. We've cleared a stream crossing and pulled out several hundred yards of road fill," she says. The steep slope has been returned to its original contour. Logs found at the bottom of the road fill have been scattered on the recontoured slope. "Down trees and decaying logs are part of the structure of old-growth forests in the Pacific Northwest. Placing the logs on the restored hillside may help migration of various fungi and other plants from the surrounding forest onto the former road."

A D-9 bulldozer, a very large piece of diesel-burning equipment, sits on the road at the end of this project area. The raw earth is claylike and clings to our shoes even though this is the dry season.

"This is the original topsoil," the geologist explains. "We look for darker-colored soil when the bulldozers are pulling the earth out of the stream crossing. When we find it, the bulldozer operator can pull it back onto the top of the recontoured incline to finish the slope. We've found that native plants are likely to sprout, probably from seed still in the soil."

In the gathering darkness, the raw slopes of the stream seem like a scene from a movie running backwards, stop-framed a moment before the bulldozers arrive to make the roadcut. The geologist continues to explain some of the lessons learned during more than a decade of experiments in the park. During 1979 and 1980 young students, some from nearby Humboldt State University, experimented with picks, shovel, hand saws, and wheelbarrows to build check dams along erosion gullies. "We believed in 'small is beautiful' and were opposed to big machines," one person who had worked on the early projects said to me. "The big bulldozers had made these roads and we thought hand tools were the most appropriate technology in the restoration of the Redwood Creek watershed. However, when I went back a decade later to sites I had worked on, I saw continuing problems of erosion. Wooden check dams we had built were rotten. New channels for erosion were emerging. I saw what the park service had done with D-9 bulldozers and I became increasingly convinced that if big equipment was used in a road-building project, big equipment was also the most practical and cost-effective way to remove it."

The restoration of lower Redwood Creek watershed in Redwood National Park is the largest project of its type ever undertaken by the Park Service. When the first generation of restoration workers began their task, Edward Abbey's novel *The Monkey Wrench Gang* was still fresh in the consciousness of many radical environmentalists. In that novel, the heroes destroy D-9 bulldozers that are being used to create an open pit mine in the Southwest.

No self-respecting radical environmentalist thought that big machines had any role in building ecotopia. But one of the lessons from a decade of restoration work in Redwood National Park is *save* some of those big, diesel-fueled bulldozers. While fossil fuel is still relatively inexpensive, use big fossil-fuel-powered equipment to rapidly remove and recontour roads, to increase the rate of healing in a landscape recovering from massive invasions by industrial civilization.

"When we use large bulldozers, we can move more earth, we can dig down through road fills, and when we find dark, decomposing material we have usually found original topsoil. After digging out the areas where fill was made for roads crossing streams, down to the original streambed or level of bedrock before road fill was made, we have less erosion than trying to build elaborate structures to channel water," the park geologist explained.

Water follows its own course. Natural processes over thousands of years created these streambeds. Geologists try to find the streambed and the topsoil taken from the road cut. They restore the landscape contours more or less as they were fifty years ago, and let nature do the replanting. There are priorities in restoration work. Restoration workers use the analogy that their work is like an emergency room in a hospital. If a patient is brought in bleeding badly, the first task is to stop the bleeding. The patient will die quickly from loss of blood. The priority in restoration of lower Redwood Creek watershed is reforming the landscape contours, slowing the erosion rate by clearing the potential erosion sites—such as log decks and stream crossings by roads. The restoration worker always has choices and has to set priorities. The priority for a restoration worker is not short-term profit, or property values, or even jobs for current generations of humans. All those are important, but they are secondary to restoration of landscape contours and giving the landscape a head start on its own healing process.

Actions undertaken during restoration work will not be aesthetically pleasing, and long-term results will not be realized until long after the person doing the work is dead. Thus it is necessary to monitor restoration projects for many years, to carefully document all work done on a restoration project so that later generations can reconstruct the sequence of change at a particular site, and to explain why certain actions were taken at certain times. Using a convoy of D-9 bulldozers in a national park may not seem appropriate to future generations. But these large machines were used to build the logging roads, and only such bulky devices have the horsepower to move large amounts of material to contour the landscape back to the form it had before logging roads were bulldozed.

Being a careful worker means being careful with the timing of actions, caring about the quality of one's work, being careful with one's attitude. Modesty is a great virtue for restoration work. Restoration workers are conscious of, and responsive to, natural events occurring on the landscape. Events are punctuation points in the flow of natural changes—hurricanes, floods, volcanic eruptions, earthquakes, landslides, ice ages, comets hitting the earth. "Normalcy" is a statistical average in an unceasing flow of natural changes. Restoration workers are aware that all their work is experimental, and aware that experimental work should be undertaken only for reasons that make sense in terms of restoring the integrity of wildness. Bureaucrats can justify horrible projects as experimental. Experimental forestry has been the justification for clearcutting ancient forests of my bioregion for a hundred years. Bureaucrats can co-opt the language, process, and vision of wildland restoration. Quality restoration work does not depend on bureaucratic needs and definition. The restoration worker relies on his or her own sensitivity to healing a particular landscape.

Restoration of wildness requires a vision of wildness. The Wildlands Project has begun to provide a method and a vision on a continental scale, but each bioregion, each small watershed, requires attention and a vision statement. The vision statement for lower Redwood Creek watershed inside the boundaries of Redwood National Park is that in five hundred years this will be mostly old-growth redwood forest. To achieve this vision will require generations of mindful attention by restoration workers. The character of restoration workers is the central factor in the human relationship with restoration of wildness. Does the restoration worker live for the life of the land? Aldo Leopold's land ethic succinctly states the meaning of restoration work: to restore the integrity, stability, and beauty of the landscape. Restoration means recovering the geography of hope. There is no escape from this planet, no aspiration to seed the stars with human creation. Our hope is here, on this planet, in the bioregion within which we dwell and survive and work.

"Think beyond your own lifetime," the geologist at Redwood National Park says, as we continue our walk through other restoration projects in the park. A generation of restoration workers can only take small steps. Institutional frameworks and culture changes are required to sustain restoration work over decades or centuries.

Restoration work is usually quiet, low profile, undramatic, without much political drama but a lot of hard labor for those doing the physical work. Most of the time the restoration worker is thinking and looking, feeling the land as the land feels the worker.

What makes a good restoration worker? "Ask why and then ask why again," says our geologist guide. Why are we doing this action? Is our motivation true to the long-term functioning of this slope? Why take out this old road at this time? Why plant these trees at this time? Why not wait a while and see what happens before rushing in where idiots have rushed before us?

Be a generalist in nature study, wildlife, fisheries, geology, natural history, sociology, ecophilosophy, and ecology. Generalists understand scientific and philosophical theories but know that all concepts and constructs are nets, and not fine enough nets to catch all the circumstances of a particular place at a particular time.

A restoration worker has a general willingness to learn from observations and to change practices based on those observations. "We made many mistakes during the first decade of this restoration project," the geologist says. "Now we have changed drastically our reasoning and the kinds of projects we undertake." The practice of restoration has a kind of openness, not a bureaucratic play-by-the-rules approach. Codifying restoration work into a manual is useful to introduce new workers to the work but can lead to a loss of creativity and lack of innovation.

In Redwood Creek, the restoration worker senses the contours of the landscape under the heavy cloak of young alder trees. The contours were altered by building logging roads and subsequent erosion after road building and logging during the past half century. A geologist does not plant trees or reintroduce top predators as a first priority. A restoration worker looks at the geomorphology, the sources of erosion and potential erosion. *Restore the soil.* That is the slogan of restoration in Redwood Creek.

A wildland restoration worker confronts, meditates on, and articulates the search for ecosophy—for wisdom to engage in what Buddhists call "right action" based on examination of our consciousness. For example, a wildland restoration worker must always address the question: is it possible to "restore wildness," or is that as much an oxymoron as "sustainable economic development"? Are we at the edge of our consciousness between modern and postmodern consciousness or only perpetuating dominant patterns of thought when we talk about "restoring wildness"? Restoration means putting everything back so that natural processes can return the area to its natural state. What does that entail in lower Redwood Creek in Redwood National Park? It entails restoring, as quickly as possible, the land contours existing before industrial logging and road building.

Priorities of restoration in Redwood National Park are somewhat different from those we hear about in The Wildlands Project. At the present point in wildland restoration, attention is directed at sustaining biodiversity. "Bring back the wolves. Bring back the predators." Many people in the wildlands movement advocate closing roads to create more wilderness. But closing a road is not the same as taking a road out. And small creatures including frogs and tree voles are important constituents of redwood forests.

Whether creating wildlife corridors, restoring wetlands, or restoring the natural flow of water in a river, wildland restoration will require confronting the roads, dams, levees, dikes, and other structures built over the last two centuries on this continent in the name of economic development. These structures are not permanent. Some already have lasted longer than anticipated. Now is the era to begin removing dams to "let rivers live." Equally important, there must be a supportive culture for restoration workers based on an ecocentric philosophy of the deep, long-range ecology movement. In such a culture each restoration worker is encouraged to explore his or her own broad identification with the ecological Self.

Each bioregional community can develop its own rituals for the withdrawal of industrial civilization into smaller parts of the landscape. When a dam is removed or a levee deleted to allow free flow of water, community celebrations with music, poetry, and speeches by politicians will be part of the process—just like the grand public rituals at openings of freeways and dams during the era of expansion in industrial civilization.

Restoration workers who incorporate a broader identification, who understand deep ecology, and who are committed to the consequences of their work for seven generations into the future are willing to support the principles of "the ecoforester's way" as stated by the Ecoforestry Institutes of the United States and Canada. These principles are stated in language that can be accepted by wilderness recovery workers, fisheries workers, forest system recovery workers, or anyone working on marine, desert, or other types of ecosystems:

1. We shall respect and learn from the ecological wisdom (ecosophy) of nature's forests with their multitudes of beings.
2. We shall protect the integrity of the full-functioning forest.
3. We shall not use agricultural practices in the forest.
4. We shall remove from the forest only values that are in abundance and that meet vital human needs.
5. We shall remove individual instances of values only when this removal will not interfere with a full-functioning forest; when in doubt, we will let them be.
6. We shall minimize the impact of our actions on the forest by using appropriate, low-impact technology.
7. We shall do good work and uphold the Ecoforester's Oath as a duty and a trust.

from *The Monarchs*

ALISON DEMING

With deep devotion, Nature, did I feel,
In that enormous City's turbulent world
Of men and things, what benefit I owed
To thee, and those domains of rural peace,
Where to the sense of beauty first my heart
Was opened . . .
    WILLIAM WORDSWORTH, *The Prelude*

But in the duration of that galactic year
we already began to realize that the world's forms
had been temporary up until then,
and that they would change, one by one.
    ITALO CALVINO, *Cosmicomics*

I

They hang in Santa Cruz by the hundreds of thousands,
shingled over each other like dead leaves
high in the eucalyptus grove,
unable to move below fifty degrees,
but getting here from everywhere west of the Rockies
in time to winter out the cold.

Their navigation takes science—an animated
scrap of paper flying two thousand miles
for the first time each year (a nine-month
life) and making it. And art to know

to move when the idea strikes. (Idea?
A butterfly idea? What could be smaller
or more frantic—yet correct. The beauties survive.)

I like to think the same intelligence,
whatever makes the monarchs fly,
is at work in my friends who shed
jobs and marriages the way a eucalyptus
does—rupturing out of its bark as it grows.

Sometimes when I'm driving I forget the car—
riding thoughts instead into
what I should have said when he said,
I guess you just don't want
to be married to anyone. Then I look up,
there's the sluice of the highway,
and I don't know how I got through the last city.

It's called *unaware memory*, what activates
when the body just does what it should.
No one remembers to accelerate,
which muscles to flex. The body
just does it because
the event is sufficiently rehearsed in the nerves.

2

What does intelligence have to do with
the mean end of Market Street
where the dying pitch tents,
unloved and contagious,
where the street screamers rail
against against against,
ones whose inside (call it "the mind")
has become the outside and the sidewalk
melts away before their pelting diatribes.

I walked by, invisible there, except
to the sorry man who approached—
"I'm not feeling too good today. Can
you help me out?" I didn't know
if he wanted my money or my mouth,
until he teased, "Oh, c'mon . . . no way?"

Safe back in my car, my blue upholstered
living room, my little local heaven . . .
stopped at the light, I saw the corner bloom
with argument—two kids on the way
home from high school, the bandannaed long-hair
pinning his girlfriend to the ground
and she screaming, "I don't want you near me!"
at first flinging the words in his face,
then, "I don't want you NEAR me!" the pitch
intensified, reaching for the businessmen
who began to slow down along the sidewalk.

One ran up, "Let her go." Nothing. Then,
"Get OFF her!" Hesitating, a thought of safety,
then lifting him like a dog by the back, the kid raging—
"Who the fuck are you! Who the FUCK are you!"
Thrashing, unable to wake up
to what had happened, still dreaming he had the right
on a street full of strangers to have his way.

3

Bypassing Santa Cruz
I descended toward Monterey Bay, a bedazzled traveler,
Connecticut Yankee ogling
the strange expanse of wealth and winterless weather.

They call it winter when the rains turn foothill grasses
velvet green, when lemons fall to the street,

when the monarchs sleep nestled in the eucalyptus leaves,
when the garlic fields quit and the lactating whales
return from the Sea of Cortez.

                                 But I grew up in the cold—
a town where, in summer, children swam in the effluent
of an explosives plant and, in winter, skated across
the surface of a dormant gray-brown world.
Call it Anywhere, U.S.A., its brick factory
now glitzed into upscale
shopping mall and all of us, the golden post-war cohort
for whom new school buildings went up
and the flag came down, half-mast, all of us
scattered like so many dandelion puffs into the urban fields.

Here on the last coast,
the tin-walled factories where herring boats used to unload
now dish out Anywhere junk by the busload. In the famous aquarium
the travelers, myself eagerly among them, marvel at how relentless
the life forms have become—pressing noses to glass
to observe one that bores through shale
to establish its niche, another with a body all
undulation and engorgement, venous, and hungry to move.

                                               Outdoors,
historic grainy photos slipping through my mind,
I recall the rubbled carcasses, remaindered bones
once lining this beach, black smoke of whale oil
and I long to know their stench because it would be
something true—those massive, black,
underwater angels harvested to light our ancestor's
dinner lamps, the hunter and the hunted now diffused and risen,
spirits in the luminous air.

4

On September 20, 1892 vast swarms
of monarchs appeared over
the city of Cleveland, reports
the *Plain Dealer*, a two-hour
flow streaming onto Superior Street
and, splendid as it was, the disordered
public mind—having just digested
news of the germ theory
of disease—mistook the visitors
for cholera germs, disguised,
of course, for the devil
assumes pleasing form. People
bundled babies, slammed windows and
tried with their newspapers
to shoo the invaders. The monarchs,
unmindful, riding a light wind,
wafted in mass undulation
from one side of the street
to the other and continued
toward their Mexican sanctuary.

5

Today the news is of monkeys,
skulls opened to expose
the neurons—brain cells, unlike
skin on starfish limbs,
unable to regenerate when they're lost—
so it matters to find out how
the message relays might re-route
to compensate for damage when

a tiny blossom unfolds
under the skull, squeezing out, say,
a person's idea of "leftness."

Though the patient might re-learn
to move the dumb hand, he doesn't
recognize it lying like a stone
in the middle of his dinner,—"Oh, yes,"
he mutters, mildly curious,
the thing no longer part of him.

In another it could be names and faces
that get lost—her husband (who she keeps
calling, "Father") bringing scrapbooks
in which she finds everyone a stranger—
parents, children, cousins, and herself.

6

Maybe the A.I. guys are right. We're
nothing more than bodies—love and
grief mere combinations
of "yes" and "no" taken to
exponential power. Nothing more

to understand than that—she's just
hardware, that woman of the nineties
who takes one slap, then grabs her
children's hands, *We're outta here*,
and they drive off into a night

so electronically bright that
the movie-goers can't forget
it's a made up world, a thousand
colored cells per second whizzing

through a lightbeam projected on the wall.
Whatever the hell this is, she prays
to a god she's pretty sure is absent
or maybe never was,
just help us to keep going.
The credits roll.

7

It must be their waking time.
Wings open to the sun,
they swing on blowing strands
of eucalyptus leaves. A few glide
and flutter in the wind.
Many have already departed,
the females landing along the way
on ripening fields of milkweed—
first in Texas, then Missouri,
Illinois—the species co-evolved
so that monarchs lay their pearly
egg-sacks on the underside of
milkweed leaves—the only meal
the young can eat, waking
into the meadow long after
the parents have departed
on their heroic flight north.
A woman lies on the railing
gazing up at the cluster of frail ones,
remembering from childhood how,
after capture, their mica-dust wings
would rub off on her hands. She nestles
the toddler onto her stomach to look up,
but the child has no patience
for a creature so small that
from this distance

she can't tell the difference
between a leaf and a wing.
But the mother thinks,
my little heroes and heroines,
in love what she thinks is a fact—
that the wind cannot unseat them
from their boughs, cannot
tumble them from flight
once the migration has begun—
in love with their flickering
awake in a little park
surrounded by ruined cities,
not a doubt in their minuscule
minds that milkweed fields await them.

8

I love the way attention
becomes a magnet. My cousin writes,
after reading of my interest
in bioluminescence,
remembering some high jinks
at the country club one night—
kids swimming to the raft
through an infestation of
phosphorescent life—glitter
falling from her arms as she stroked
through the dark silk of water.

The marina onshore spit its
puny artificial light at them
as the pack of startled teenagers
clambered onto the raft to rest,
the churn of their excitement

making the water slap and
gong inside the metal floats.
And when the noise had settled
into lapping, each one began

to listen hard. They might
have been castaways, citizens
only of the wild biology
surrounding them, the sheen
of an unknown metabolism
taunting them to cut
the anchor lines and figure out
from scratch how to stay alive.

The Contributors

D AVID A BRAM is an ecologist and philosopher whose essays have appeared in *Orion*, *Wild Earth*, *Environmental Ethics*, *The Ecologist*, *Utne Reader*, *Parabola*, and numerous anthologies. An avid student of interspecies communication, he has lived and studied among indigenous peoples in Indonesia, Nepal, and North America. His forthcoming book, *Inconceivable Earth: Animism, Language, and the Ecology of Sensory Experience*, will be published in 1995.

D AVID C LARKE B URKS is a writer, editor, and lecturer in the Environmental Studies Program at the University of Oregon. His articles, essays, and stories have appeared in various publications across the country, and he is a correspondent for *Wild Earth*. He has traveled extensively throughout North and Central America and is currently at work on a collection of essays.

PHYLLIS R ACHAEL B URKS is an illustrator and graphic designer living in the Pacific Northwest. Her illustrations have appeared in *Wild Earth* and *Forest Voice*.

J OHN D AVIS is editor of *Wild Earth*, board member of The Wildlands Project, and conservation codirector with Jamie Sayen for the Greater Laurentian Region of The Wildlands Project. Formerly he edited *Earth First! Journal*, and he compiled and edited *The Earth First! Reader*. He lives in the Green Mountains but considers the much wilder Adirondacks his home.

A LISON D EMING is director of the University of Arizona Poetry Center. Her collection, *Science and Other Poems*, won the Walt Whitman Award. A collection of nature essays, *Temporary Homelands*, is slated for publication by August 1994.

B ILL D EVALL is professor of sociology at Humboldt State University in Arcata, California. He is the author of numerous books and articles, as well as a leading voice for the deep ecology movement. He is the author of *Simple in*

Means, Rich in Ends: Living Richly in an Age of Limits and coauthor of *Deep Ecology: Living as If Nature Mattered.*

ALAN DRENGSON is an associate professor of philosophy at the University of Victoria, B.C., Canada. He is the founder and editor of *The Trumpeter* and author of *Beyond Environmental Crisis.*

PAUL FAULSTICH teaches interdisciplinary environmental studies at Pitzer College in Claremont, California. His passions lie at the intersection of culture and nature: Fourth World political ecology, totemic geography, and the ecology of expressive culture. He is a correspondent for *Wild Earth* and other journals and an advocate for ecological and cultural diversity.

DAVE FOREMAN is chairman of The Wildlands Project and executive editor of *Wild Earth.* He was the Wilderness Society's southwest regional representative from 1973 to 1977 and 1979 to 1980 and the society's director of wilderness affairs in Washington, D.C., from 1977 to 1980. He has been an activist with the New Mexico Wilderness Study Committee, Sierra Club, Nature Conservancy, Earth First!, and many other conservation groups. He is the author of *Confessions of an Eco-Warrior* and *The Big Outside* (with Howie Wolke) and owns a mail-order conservation bookstore.

MITCH FRIEDMAN is executive director of the Greater Ecosystem Alliance, Bellingham, Washington. He has worked extensively to apply landscape ecology and conservation biology to environmental advocacy. Among his productions and publications are a videotape, *Biodiversity Is the Variety of Life,* and a book, *Cascadia Wild: Conserving an International Ecosystem.* He serves on the board of The Wildlands Project and several other conservation groups.

ED GRUMBINE is the author of *Ghost Bears* and directs the Sierra Institute, a wildlands studies undergraduate program of the University of California Extension, Santa Cruz. His primary interests include bringing deep ecological values into mainstream environmental policy, the relationship between wildness and people, and the lay practice of natural history.

JOHN HAINES is one of Alaska's most esteemed poets and essayists. Among his published works are *Winter News, The Stone Harp, Cicada, News from the Glacier: Selected Poems 1960–1980,* and two collections of essays. His collected poems, *The Owl in the Mask of the Dreamer,* published in 1993, were awarded the William Stafford Memorial Award by the Pacific Northwest Booksellers Association, and a new volume of essays is forthcoming in 1994.

DAVID JOHNS is executive director of The Wildlands Project and teaches courses on politics and conservation. He has written and spoken widely on integrating science, activism, and ecocentric values. Close to his home he has worked to protect Oregon's high desert from mining and industrial cattle production; he has also planted trees in Nicaragua and worked to protect grizzlies.

DOLORES LACHAPELLE is a teacher of tai chi, Taoist philosophy, deep ecology, primitive rituals, deep powder skiing, and other strains of earth wisdom. She is the author of *Sacred Land, Sacred Sex*; *Earth Wisdom*; and, most recently, *Deep Powder Snow: 40 Years of Ecstatic Skiing, Avalanches, and Earth Wisdom*.

NANCY LORD lives in Homer, Alaska, and is the author of two collections of short stories, including *Survival*. She is currently working on a nonfiction book about the human and natural history around her summer fish camp in Alaska.

CHRISTOPHER MANES is an attorney with a law firm in Palm Springs, California. He is the author of *Green Rage: Radical Environmentalism and the Unmaking of Civilization*, which was nominated for a 1990 L.A. Times book award in science. His articles have appeared in *Lear's*, *Environmental Ethics*, *Orion*, *Penthouse*, *Northern Lights*, and elsewhere.

MOLLIE YONEKO MATTESON, a correspondent for *Wild Earth*, first found wildness tromping the woods of Vermont. She researched wolves in Montana for an MS in wildlife biology and has worked as a wilderness ranger, biologist, environmental educator, and wolf advocate.

BILL MCKIBBEN is an Adirondack resident, advocate, and writer. He is the author of *The End of Nature* and *The Age of Missing Information*.

MONIQUE A. MILLER is executive director of Carrying Capacity Network (CCN). She also serves as editor of CCN's two publications: the monthly newsletter *Clearinghouse Bulletin* and the quarterly journal *Focus*. She speaks and writes extensively about U.S. population growth and environmental degradation issues.

STEPHANIE MILLS is a writer and bioregional activist living on the Leelanau Peninsula of Lower Michigan. She served as editor of *Not Man Apart*, *Co-Evolution Quarterly*, *California Tomorrow*, and *In Praise of Nature*. She is the author of *Whatever Happened to Ecology?* (1989). Her most recent book is about ecological restoration.

JAMES MORRISON is an ethnohistorian and heritage consultant with extensive experience in aboriginal claims research in Canada. A resident of Haileybury in northeastern Ontario, he was an executive member of the Lady Evelyn Wilderness Alliance, which lobbied in the early 1980s for the creation of the Lady Evelyn–Smoothwater Wilderness Park. Since August 1992, he has been chair of the Temagami Area Wendaban Stewardship Authority.

GARY PAUL NABHAN is writer-in-residence at the Arizona Sonora Desert Museum and a cofounder of Native Seeds. His latest books are *Counting Sheep* and *The Geography of Childhood* (with Steve Trimble). He has received a MacArthur Award and the John Burroughs medal for nature writing.

REED F. NOSS is the editor of *Conservation Biology*, science director of The Wildlands Project, research scientist at the University of Idaho, and an adjunct associate professor at Oregon State University. He has worked with the Ohio Department of Natural Resources, the Nature Conservancy, EPA, and others during his twenty-two years in the environmental field. An author of over seventy-five publications, he lives with his family in the foothills of the Coast Range near Corvallis, Oregon.

MAX OELSCHLAEGER teaches graduate courses in the philosophy of ecology, postmodern thought, and environmental ethics at the University of North Texas. His books include *The Idea of Wilderness*, *Caring for Creation*, and *The Environmental Imperative*. He has edited *The Wilderness Condition*, *After Earth Day*, and *Postmodern Environmental Ethics*, and his essays have appeared in *Environmental Ethics*, *The Journal of Social Philosophy*, *Wild Earth*, *Ekistics*, *Southwest Philosophical Studies*, and elsewhere.

KIRKPATRICK SALE is the author of seven books, including *Conquest of Paradise: Christopher Columbus and the Columbian Legacy* and *The Green Revolution: American Environment Movement, 1962–92*. He is at work on a book about the Luddites, *Rebels Against the Future: Lessons from the First Industrial Revolution*.

JAMIE SAYEN is the editor of *Northern Forest Forum*, director of the Northern Appalachian Restoration Project of Earth Island Institute, and a board member of The Wildlands Project. He lives in a log cabin just beyond Cross-Eyed Corners in Stratford Hollow, New Hampshire.

GARY SNYDER has lived and traveled in wildlands all around the Pacific Rim. He got his start working on the farm and climbing snowpeaks in the Pacific Northwest. He now lives in the northern Sierra Nevada and works on ecological and social issues with his local Yuba Watershed Institute. He also teaches part-time at the University of California at Davis. His latest books are *Practice of the Wild* (prose), *No Nature* (selected poems), and *Coming Into the Watershed* (new and selected prose, due out in 1995).

JACK TURNER has been climbing mountains for thirty-three years. He has taught philosophy at the University of Illinois and for the past fifteen years has lived in Grand Teton National Park where he is chief guide for the Exum Guide Service and School of Mountaineering. His collection of essays, *The Abstract Wild*, is forthcoming.

JAY HANSFORD C. VEST is assistant professor of Native American cultures and literatures at Arizona State University West in Phoenix. His interests include the philosophy of ecology and comparative mythology.

TERRY TEMPEST WILLIAMS is naturalist-in-residence at the Utah Museum of Natural History. Her most recent book is *Refuge: An Unnatural History of Family and Place*. She currently serves on the governing council of The Wilderness Society. She is the recipient of a nonfiction fellowship from the Lannan Foundation.

GEORGE WUERTHNER, a member of The Wildlands Project board of directors, is an ecologist, wilderness explorer, photographer, and freelance writer. Author of fifteen books and countless articles, he has traveled extensively throughout North America and visited several hundred designated wilderness areas in the United States.

MARGARET HAYS YOUNG is a writer, actor, and full-time volunteer leader with Alliance for the Wild Rockies, Preserve Appalachian Wilderness, Sierra Club Atlantic Chapter, and other groups.

Grateful acknowledgment is expressed for permission to include the following previously published material.

"The Far Outside" by Gary Paul Nabhan is adapted from a talk given at "Art of the Wild," Squaw Valley Writer's Conference, Squaw Valley, CA. Summer 1993. Used by permission of Victoria Shoemaker, Literary Agent.

"Water Songs" by Terry Tempest Williams is from *An Unspoken Hunger: Stories from the Field*. Copyright © 1994 by Terry Tempest Williams. Reprinted by permission of the author and Pantheon Books, a division of Random House, Inc.

"The Wild and the Tame" by Stephanie Mills is from a forthcoming work, *The Art and Science of Land Restoration*, by the author. Copyright © 1995 by Stephanie Mills. Reprinted by permission of the author and Beacon Press.

"Sacred Geography of the Pikuni: The Badger–Two Medicine Wildlands" by Jay Hansford C. Vest, *Western Wildlands*, Fall 1989, 15(3). Reprinted by permission of the author and the publisher.

"Scattered Notes on the Relation Between Language and the Land" by David Abram is from a forthcoming work, *Inconceivable Earth*, by the author (New York: Pantheon Books). Used by permission of the author.

"Notes from an Interrupted Journal" by John Haines, *The Ohio Review*, 1988, No. 47. Reprinted by permission of the author and the publisher.

"The Rhizome Connection" by Dolores LaChapelle, *Wild Earth*, Fall 1993. Reprinted by permission of the author and the publisher, P.O. Box 455, Richmond, VT 05477.

"Covers the Ground" by Gary Snyder is from work in progress entitled *Mountains and Rivers Without End*. Used by permission of the poet.

"Protected Areas and Aboriginal Interests in Canada" by James Morrison, July 1993, paper published by World Wildlife Fund Canada. Reprinted by permission of author and publisher.

Poems 1 through 8 from *The Monarchs* by Alison Deming. Used by permission of the poet. Poems 1 and 3 appeared previously in *The Eloquent Edge: 15 Maine Women Writers* (Bar Harbor, Me.: Acadia Publishing Company) and are reprinted by permission of the publisher.

Index